EARTH SCIENCES IN THE 21ST CENTURY

RARE EARTH ELEMENTS IN THE U.S. AND CHINA

EARTH SCIENCES IN THE 21ST CENTURY

Additional books in this series can be found on Nova's website under the Series tab.

Additional E-books in this series can be found on Nova's website under the E-books tab.

EARTH SCIENCES IN THE 21ST CENTURY

RARE EARTH ELEMENTS IN THE U.S. AND CHINA

JOSEPH A. MEYER
AND
DANIEL M. RODRIGUES
EDITORS

Nova Science Publishers, Inc.
New York

For permission to use material from this book please contact us:
Telephone 631-231-7269; Fax 631-231-8175
Web Site: http://www.novapublishers.com

NOTICE TO THE READER

Additional color graphics may be available in the e-book version of this book.

Library of Congress Cataloging-in-Publication Data

Rare earth elements in the U.S. and China / editors, Joseph A. Meyer and Daniel M. Rodrigues.
p. cm.
Includes index.
ISBN 978-1-61324-500-2 (hardcover)
1. Rare earth metals--United States. 2. Rare earth metals--China. I. Meyer, Joseph A. (Joseph Aaron), 1972- II. Rodrigues, Daniel M. (Daniel Mateo), 1974-
TN490.A2R365 2011
381'.42494--dc23

2011014212

Published by Nova Science Publishers, Inc. † New York

CONTENTS

PREFACE

Rare earth elements (REE) are essential for a diverse and expanding array of high-technology applications. REE-containing magnets, metal alloys for batteries and light-weight structures, and phosphors are essential for many current and emerging alternative energy technologies, such as electric vehicles, energy-efficient lighting and wind power. REE are also critical for a number of key defense systems and other advanced materials. This book provides a nontechnical overview of domestic reserves and resources of REE and possibilities for utilizing those resources. At the present time, the U.S. obtains its REE raw materials from foreign sources, almost exclusively from China.

Chapter 1- The rare earth elements (REE) are fifteen elements with atomic numbers 57 through 71, from lanthanum to lutetium ("lanthanides"), plus yttrium (39), which is chemically similar to the lanthanide elements and thus typically included with the rare earth elements. Although industrial demand for these elements is relatively small in tonnage terms, they are essential for a diverse and expanding array of high-technology applications. REE-containing magnets, metal alloys for batteries and light-weight structures, and phosphors are essential for many current and emerging alternative energy technologies, such as electric vehicles, energy-efficient lighting, and wind power. REE are also critical for a number of key defense systems and other advanced materials.

Chapter 2- China's dominant position as the producer of over 95 percent of the world output of rare-earth minerals and rapid increases in the consumption of rare earths owing to the emergence of new clean- energy and defense-related technologies, combined with China's decisions to restrict exports of rare earths, have resulted in heightened concerns about the future availability of rare earths. As a result, industrial countries such as Japan, the United States, and countries of the European Union face tighter supplies and higher prices for rare earths. This paper briefly reviews China's rare-earth production, consumption, and reserves and the important policies and regulations regarding the production and trade of rare earths, including recently announced export quotas.

In: Rare Earth Elements in the U.S. and China
Editors: Joseph A. Meyer et al.
ISBN: 978-1-61324-500-2

Chapter 1

The Principal Rare Earth Elements Deposits of the United States: A Summary of Domestic Deposits and a Global Perspective

U.S. Geological Survey

Abbreviations Used in This Report

cm	centimeter
ft	foot
ft^3	cubic foot
g/cm^3	gram per cubic centimeter
in.	inch
km	kilometer
km^2	square kilometer
kt	thousand metric tons
lb	pound
m	meter
mi	mile
mi^2	square mile
mm	millimeter
ppm	parts per million

t	metric ton
HREE	heavy rare earth elements
LREE	light rare earth elements
REE	rare earth elements
U.S.	United States
USGS	United States Geological Survey
REO	rare earth oxide
TREO	total rare earth oxide
WGS84	World Geodetic System of 1984 (the reference coordinate system used by global positioning systems)

INTRODUCTION AND BACKGROUND

The rare earth elements (REE) are fifteen elements with atomic numbers 57 through 71, from lanthanum to lutetium ("lanthanides"), plus yttrium (39), which is chemically similar to the lanthanide elements and thus typically included with the rare earth elements. Although industrial demand for these elements is relatively small in tonnage terms, they are essential for a diverse and expanding array of high-technology applications. REE-containing magnets, metal alloys for batteries and light-weight structures, and phosphors are essential for many current and emerging alternative energy technologies, such as electric vehicles, energy-efficient lighting, and wind power. REE are also critical for a number of key defense systems and other advanced materials.

Section 843 of the National Defense Authorization Act for Fiscal Year 2010, Public Law 111-84, directs the Comptroller General to complete a report on REE materials in the defense supply chain. The Office of Industrial Policy, in collaboration with other U.S. Government agencies, has initiated (in addition to this report) a detailed study of REE. This latter study will assess the Department of Defense's use of REE, as well as the status and security of domestic and global supply chains. That study will also address vulnerabilities in the supply chain and recommend ways to mitigate any potential risks of supply disruption. To help conduct this study, the Office of Industrial Policy asked the U.S. Geological Survey (USGS) to report on domestic REE reserves and resources in a global context. To this end, the enclosed report is the initial USGS contribution to assessing and summarizing the domestic REE resources in a global perspective.

In 2009, the Mineral Resources Program of the USGS organized a new project under the title Minerals at Risk and For Emerging Technologies in order to evaluate mineral resource and supply issues of rare metals that are of increasing importance to the national economy. Leaders and members of this project, with the assistance of the USGS National Minerals Information Center, prepared the enclosed USGS report on domestic REE resources. The USGS Mineral Resources Program has investigated domestic and selected foreign REE resources for many decades, and this report summarizes what has been learned from this research. The USGS National Minerals Information Center (formerly Minerals Information Team) has monitored global production, trade, and resources for an equally long period and is the principal source of statistics used in this report.

The objective of this study is to provide a nontechnical overview of domestic reserves and resources of REE and possibilities for utilizing those resources. At the present time, the United States obtains its REE raw materials from foreign sources, almost exclusively from China. Import dependence upon a single country raises serious issues of supply security. In a global context, domestic REE resources are modest and of uncertain value; hence, available resources in traditional trading partners (such as Canada and Australia) are of great interest for diversifying sources of supply. This report restates basic geologic facts about REE relevant to assessing security of supply, followed by a review of current United States consumption and imports of REE, current knowledge of domestic resources, and possibilities for future domestic production. Further detail follows in a deposit-by-deposit review of the most significant domestic REE deposits (see index map). Necessary steps to develop domestic resources are discussed in a separate section, leading into a review of current domestic exploration and a discussion of the value of a future national mineral resource assessment of REE. The report also includes an overview of known global REE resources and discusses the reliability of alternative foreign sources of REE.

The Rare Earth Elements

The rare earth elements (REE) comprise 15 elements that range in atomic number from 57 (lanthanum) to 71 (lutetium) on the periodic table (figure 1). These elements are also commonly referred to as "lanthanides." Yttrium (atomic number = 39) is also included with the REE group, because it shares chemical and physical similarities with the lanthanides.

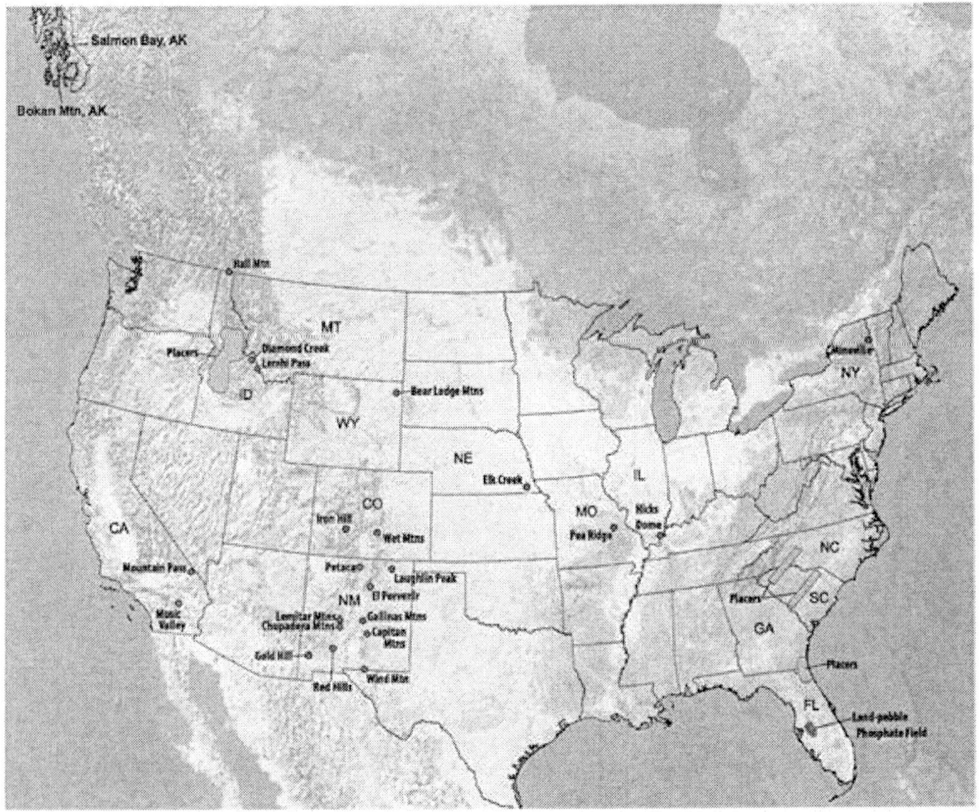

Principal rare earth elements districts in the United States, which are described in this report.

Traditionally, the REE are divided into two groups on the basis of atomic weight: the light rare earth elements are lanthanum through europium (atomic numbers = 57 through 63); and the heavy rare earth elements are gadolinium through lutetium (atomic numbers = 64 through 71). Yttrium (atomic number = 39), although light, is included with the heavy REE group because of its common chemical and physical affiliations with the heavy REE in nature.

Most of the REE are not as rare as the group's name suggests. They were named rare earth elements because most were identified during the 18th and 19th centuries as oxide components within seemingly rare minerals. Cerium is the most abundant REE, and it is actually more common in the Earth's crust than is copper or lead. All of the REE except promethium are more abundant than silver or mercury (Taylor and McLennan, 1985). The rare earth elements are commonly found together in the Earth's crust because they share a trivalent charge ($^{+3}$) and similar ionic radii. Detailed information on the REE is described in Emsley (2001), and an overview of the geology, production, and economics of REE is provided by Castor and Hedrick (2006).

Basic Geology of Rare Earth Elements

Several geologic aspects of the natural occurrence of rare earth elements strongly influence the supply of rare-earth- elements raw materials. These geologic factors are presented as statements of facts followed by a detailed discussion. This section is placed before the balance of the report because an understanding of these facts is critical to the discussion that follows and should be read first.

Although rare earth elements are relatively abundant in the Earth's crust, they are rarely concentrated into mineable ore deposits.

The estimated average concentration of the rare earth elements in the Earth's crust, which ranges from around 150 to 220 parts per million (table 1), exceeds that of many other metals that are mined on an industrial scale, such as copper (55 parts per million) and zinc (70 parts per million). Unlike most commercially mined base and precious metals, however, rare earth elements are rarely concentrated into mineable ore deposits. The principal concentrations of rare earth elements are associated with uncommon varieties of igneous rocks, namely alkaline rocks and carbonatites. Potentially useful concentrations of REE-bearing minerals are also found in placer deposits, residual deposits formed from deep weathering of igneous rocks, pegmatites, iron-oxide copper-gold deposits, and marine phosphates (table 2).

Alkaline igneous rocks form from cooling of magmas derived by small degrees of partial melting of rocks in the Earth's mantle. The formation of alkaline rocks is complex and not fully understood but can be thought of as a geologic process that extracts and concentrates those elements that do not fit into the structure of the common rock-forming minerals. The resulting alkaline magmas are rare and unusually enriched in elements such as zirconium, niobium, strontium, barium, lithium, and the rare earth elements. When these magmas ascend into the Earth's crust, their chemical composition undergoes further changes in response to variations in pressure, temperature, and composition of surrounding rocks. The result is an astonishing diversity of rock types that are variably enriched in economic elements, including the rare earth elements. The mineral deposits associated with these rocks are likewise quite diverse and awkward to classify, in that the distinctive features of these deposits and their rarity can result in classifications that have only one or a few known examples.

Classification of ores related to alkaline rocks is also controversial. Table 2 presents a relatively simple classification that follows analogous categories for deposits related to nonalkaline igneous rocks. Some of the more unusual alkaline rocks that host, or are related to, REE ores are carbonatite and phoscorite, igneous rocks composed principally of carbonate and phosphate minerals, respectively. Carbonatites, and especially phoscorites, are relatively uncommon, as there are only 527 known carbonatites in the world (Woolley and Kjarsgaard, 2008). Economic concentrations of REE-bearing minerals occur in some alkaline rocks, skarns and carbonate-replacement deposits associated with alkaline intrusions, veins and dikes cutting alkaline igneous complexes and surrounding rocks, and soils and other weathering products of alkaline rocks.

Weathering of all types of rocks yields sediments that are deposited in a wide variety of environments, such as streams and rivers, shorelines, alluvial fans, and deltas. The process of erosion concentrates denser minerals, most notably gold, into deposits known as placers. Depending on the source of the erosion products, certain rare earth elements–bearing minerals, such as monazite and xenotime, can be concentrated along with other heavy minerals. The source need not be an alkaline igneous rock or a related rare-earth deposit. Many common igneous, metamorphic, and even older sedimentary rocks contain enough monazite to produce a monazite-bearing placer. As a result, monazite is almost always found in any placer deposit. However, the types of placers with the greatest concentrations of monazite are typically ilmenite–heavy mineral placers, which have been mined for titanium oxide pigments, and cassiterite placers, which are mined for tin.

In tropical environments, rocks are deeply weathered to form a unique soil profile consisting of laterite, an iron- and aluminum-rich soil, as much as many tens of meters thick. The processes of soil formation commonly concentrate heavy minerals as residual deposits, resulting in an enriched-metal layer over the underlying, unweathered bedrock. When a rare- earth deposit undergoes such weathering, it may be enriched in rare earth elements in concentrations of economic interest.

A particular type of REE deposit, the ion-absorption type, is formed by the leaching of rare earth elements from seemingly common igneous rocks and fixing the elements onto clays in soil. These deposits are only known in southern China and Kazakhstan and their formation is poorly understood.

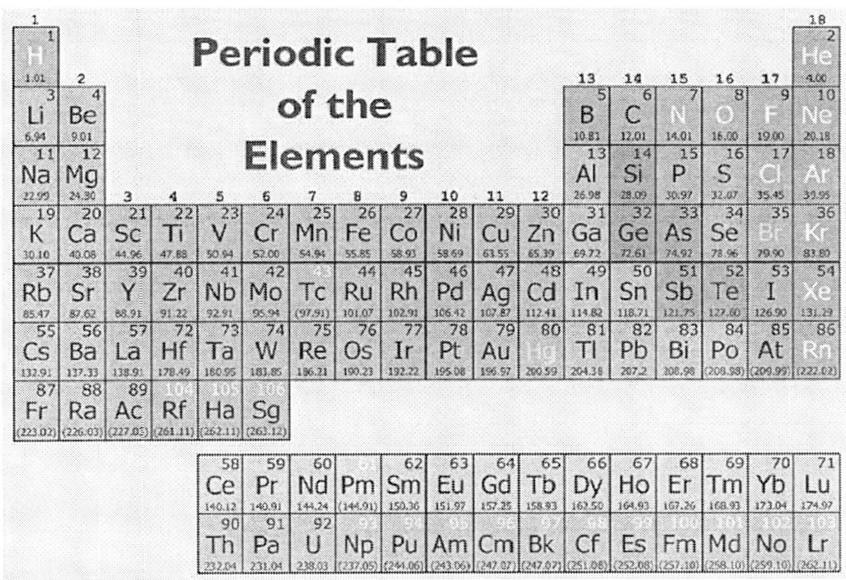

Figure 1. Periodic table of the elements. The rare earth elements comprise 15 elements, which range in atomic number from 57 to 71, including lanthanum (La) to lutetium (Lu). The elements are also commonly referred to as "lanthanides." Yttrium (Y, atomic number 39) is also typically included with the rare earth elements group because it shares chemical, physical, and application properties with the lanthanides.

Table 1. Estimates of the Crustal Abundances of Rare Earth Elements
[Rare earth elements listed in order of increasing atomic number; yttrium (Y) is included with these elements because it shares chemical and physical similarities with the lanthanides. Unit of measure, parts per million]

Rare earth element	Mason and Moore (1982)	Jackson and Christiansen (1993)	Sabot and Maestro (1995)	Wedephol (1995)	Lide (1997)	McGill (1997)
Lanthanum	30	29	18	30	39	5 to 18
Cerium	60	70	46	60	66.5	20 to 46
Praseodynium	8.2	9	5.5	6.7	9.2	3.5 to 5.5
Neodymium	28	37	24	27	41.5	12 to 24
Samarium	6	8	6.5	5.3	7.05	4.5 to 7
Europium	1.2	1.3	0.5	1.3	2	0.14 to 1.1
Gadolinium	5.4	8	6.4	4	6.2	4.5 to 6.4
Terbium	0.9	2.5	0.9	0.65	1.2	0.7 to 1
Dysprosium	3	5	5	3.8	5.2	4.5 to 7.5
Holmium	1.2	1.7	1.2	0.8	1.3	0.7 to 1.2
Erbium	2.8	3.3	4	2.1	3.5	2.5 to 6.5
Thulium	0.5	0.27	0.4	0.3	0.52	0.2 to 1
Ytterbium	3.4	0.33	2.7	2	3.2	2.7 to 8
Lutetium	0.5	0.8	0.8	0.35	0.8	0.8 to 1.7
Yttrium	33	29	28	24	33	28 to 70
Scandium	22		10	16	22	5 to 10
Total	206.1	205.2	159.9	184.3	242.17	

Among pegmatites, a group of very coarse grained intrusive igneous rocks, the niobium-yttrium-fluorine family, comprises a large number of subtypes formed in different geologic environments. These subtypes are granitic in composition and are usually found peripheral to large granitic intrusions. In general, however, rare earth elements–bearing pegmatites are generally small and are of economic interest only to mineral collectors.

The iron-oxide copper-gold type of deposit has been recognized as a distinct deposit type only since the discovery of the giant Olympic Dam deposit in South Australia in the 1980s. The Olympic Dam deposit is unusual in that it contains large amounts of rare earth elements and uranium. An economic method for recovering rare earth elements from these deposits has not yet been found. Many other deposits of this type have been identified around the world, but information on their rare earth elements content is generally lacking. Trace amounts of rare earth elements have also been identified in magnetite-apatite replacement deposits.

Karst bauxites, aluminum-rich soils that accumulate in cavernous limestone (underlying karst topography) in Montenegro and elsewhere, are enriched in rare earth elements, but the resulting concentrations are not of economic interest (Maksimović and Pantó, 1996). The same can be said for marine phosphate deposits, which can contain as many as 0.1 percent REE oxides (Altschuler and others, 1966). As a result, recovery of rare earth elements as a byproduct of phosphate fertilizer manufacture has been investigated.

The ores of rare earth elements are mineralogically and chemically complex and commonly radioactive.

In many base and precious metal deposits, the metals extracted are highly concentrated in a single mineral phase, such as copper in chalcopyrite ($CuFeS_2$) or zinc in sphalerite (ZnS). Separation of a single mineral phase from rock is a relatively easy task. The final product is a concentrate typically sent to a smelter for final extraction and refining of the metals. Zinc, for example, is almost entirely derived from the mineral sphalerite, such that the global zinc smelting and refining industry has developed a highly specialized treatment of this mineral. Thus, production of zinc has a pronounced cost advantage in that a single standard technology is used, and the development of a new zinc mine is a largely conventional process.

Current mineral-processing practice is capable of sequential separation of multiple mineral phases but it is not always cost effective to do so. When elements of interest are found in two or more mineral phases, each requiring a different extraction technology, mineral processing is relatively costly. Many rare earth elements deposits contain two or more rare earth elements–bearing phases. Therefore, rare earth elements deposits in which the rare earth elements are largely concentrated in a single mineral phase have a competitive advantage. To date, REE production has largely come from single-mineral-phase deposits, such as Bayan Obo (bastnasite), Mountain Pass (bastnasite), and heavy-mineral placers (monazite).

Table 2. Classification of Rare Earth Elements–Bearing Mineral Deposits

Association	Type	Example
Peralkaline igneous rocks	Magmatic (alkali-ultrabasic)	Lovozero, Russia.
	Pegmatite dikes (alkali-ultrabasic)	Khibina Massif, Russia.
	Pegmatite dikes (peralkaline)	Motzfeldt, Greenland.
	Hydrothermal veins and stockworks	Lemhi Pass, Idaho.
	Volcanic	Brockman, Western Australia.
	Metasomatic-albitite	Miask, Russia.
Carbonatites	Magmatic	Mountain Pass, California.
	Dikes and dialational veins	Kangakunde Hill, Malawi.
	Hydrothermal veins and stockworks	Gallinas Mtns., New Mexico.
	Skarn	Saima, China.
	Carbonate rock replacement	Bayan Obo, China.
	Metasomatic-fenite	Magnet Cove, Arkansas.
Iron oxide copper-gold	Magnetite-apatite replacement	Eagle Mountain, California.
	Hematite-magnetite breccia	Olympic Dam, South Australia.
Pegmatites	Abyssal (heavy rare earth elements)	Aldan, Russia.
	Abyssal (light rare earth elements)	Five Mile, Ontario.
	Muscovite (rare earth elements)	Spruce Pine, North Carolina.
	Rare earth elements-allanite-monazite	South Platte, Colorado.
	Rare earth elements-euxenite	Topsham, Maine.
	Rare earth elements-gadolinite	Ytterby, Sweden.
	Miarolitic-rare earth elements-topaz-beryl	Mount Antero, Colorado
	Miarolitic-rare earth elements-gadolinite-fergusonite	Wasau complex, Wisconsin.
Porphyry molybdenum	Climax-type	Climax, Colorado.
Metamorphic	Migmatized gneiss	Music Valley, California.
	Uranium–rare earth elements skarn	Mary Kathleen, Queensland.
Stratiform phosphate residual	Platform phosphorite	Southeast Idaho.
	Carbonatite-associated	Mount Weld, Western Australia.
	Granite-associated laterite	South China.
	Baddeleyite bauxite	Poços de Caldas, Brazil.
	Karst bauxite	Montenegro.
Paleoplacer	Uraniferous pyritic quartz pebble conglomerate	Elliot Lake, Ontario.
	Auriferous pyritic quartz pebble conglomerate	Witwatersrand, South Africa.
Placer	Shoreline Ti–heavy mineral placer	Cooljarloo, Western Australia.
	Tin stream placer	Malaysia.

Rare earth elements–bearing minerals, once separated, contain as many as 14 individual rare earth elements (lanthanides and yttrium) that must be further separated and refined. The complexity of extracting and refining rare earth elements is illustrated by a metallurgical flow sheet for the Mountain Pass mine in California (figure 2). Unlike metal sulfides, which are chemically simple compounds, REE-bearing minerals are quite complex (table 3). Base metal sulfide ores, such as sphalerite (ZnS), are typically smelted to burn off sulfur and separate impurities from the molten metal. The resulting metal is further refined to near purity by electrolysis. Rare earth elements, on the other hand, are typically extracted and refined through dozens of chemical processes to separate the different rare earth elements and remove impurities.

The principal deleterious impurity in REE-bearing minerals is thorium, which imparts an unwanted radioactivity to the ores. Because radioactive materials are difficult to mine and handle safely, they are heavily regulated. When a radioactive waste product is produced, special disposal methods must be used. The cost of handling and disposing of radioactive material is a serious impediment to the economic extraction of the more radioactive REE-rich minerals, in particular monazite, which typically contains considerable amounts of thorium. In fact, imposition of tighter regulations on the use of radioactive minerals drove many sources of monazite out of the rare earth elements market during the 1980s.

The complex metallurgy of rare earth elements is compounded by the fact that no two REE ores are truly alike. As a result, there is no standard process for extracting the REEbearing minerals and refining them into marketable rare earth compounds. To develop a new rare earth elements mine, the ores must be extensively tested by using a variety of known extraction methods and a unique sequence of optimized processing steps. Compared with a new zinc mine, process development for rare earth elements costs substantially more time and money.

Mineralogy of United States Deposits

The main REE-bearing minerals found in the United States are euxenite, bastnasite, xenotime, monazite, and allanite. Samarskite, aeschynite, fergusonite, parisite, synchisite, tengerite, ancylite, florencite, britholite, kainosite, and thalenite have also been identified in United States deposits (table 4). Euxenite $[(Y,Er,Ce,U,Pb,Ca)(Nb,Ta,Ti)_2(O,OH)_6]$ is an oxide mineral that forms a series with the mineral polycrase $[(Y,Ca,Ce,U,Th)(Ti,Nb,Ta)_2O_6]$. Other rare earth elements oxide (REO) minerals, such as fergusonite, aeschynite, and samarskite, have similar properties, making identification in hand sample difficult. Euxenite is black with a tabular to prismatic habit, making it indistinguishable from rutile, a common oxide mineral, when found in massive form. However, euxenite does not have any cleavage planes and, unlike rutile, has a conchoidal fracture. Furthermore, because euxenite is ordinarily found in granite pegmatites, it is commonly associated with quartz, feldspars, columbite (now called ferrocolumbite), tantalite (now called ferrotantalite or manganocolumbite), and monazite.

Bastnasite (also spelled bastnäsite or bastnaesite) is a rare REE-bearing carbonate mineral [$(Ce, La,Y)CO_3F$] that forms a series with the mineral hydroxyl-bastnasite [$(Ce,La)CO_3(OH,F)$]. Bastnasite can be pale white, tan, gray, brown, yellow, or pink, with a pearly, vitreous, or greasy to dull luster. Bastnasite usually forms small rounded hexagonal or short prismatic crystals, though it can also form rosettes and spheres. Both massive and granular varieties have been observed. Bastnasite is closely related to the mineral parisite [$Ca(Ce,La)_2(CO_3)_3F_2$] and has been known to replace crystals of allanite.

Xenotime is a Y-bearing phosphate mineral (YPO_4) and can be yellowish brown to reddish brown with a vitreous to resinous luster. Less common colors include gray, salmon pink, and green. Xenotime is usually an accessory mineral in acidic and alkaline rocks, though it has been observed in mica schists and quartz-rich gneisses; it may also be a detrital mineral. Xenotime can easily be confused with zircon because of similarities in crystal habit and overall appearance. However, xenotime is not as hard as zircon and demonstrates perfect {100} cleavage.

Monazite is a REE- and thorium-bearing phosphate mineral [$(Ce,La,Y,Th)PO_4$] and typically contains 60–62 percent total rare-earth oxides. Monazite's resistance to chemical weathering and its high specific gravity account for its association with other resistant heavy minerals such as ilmenite, magnetite, rutile, and zircon. Because monazite is radioactive, however, grains can be metamict, which means they have lost their crystalline structure owing to radioactive decay. Crystals of monazite are yellow to brown or orange-brown with a vitreous and resinous or adamantine luster. Monazite grains are usually equant to prismatic with wedge-shaped terminations. Both granular and massive forms exist.

Allanite [$Ca(Ce,La,Y,Ca)Al_2(Fe^{2+},Fe^{3+})(SiO_4)(Si_2O_7)\ O(OH)$], which belongs to the epidote mineral group, is one of the more common REE-bearing minerals in igneous rocks but is rarely concentrated enough to be an ore of REE. Allanite grains are tabular and usually black, though dark brown to brownish violet varieties also occur. Allanite has a conchoidal fracture and is commonly metamict because of the radioactive decay of thorium. The presence of a halo or dark ring inside the mineral grain is also an effect of its radioactivity. Most commonly, allanite is found as an accessory mineral in igneous rocks, such as granites, syenites, diorites, and associated pegmatites.

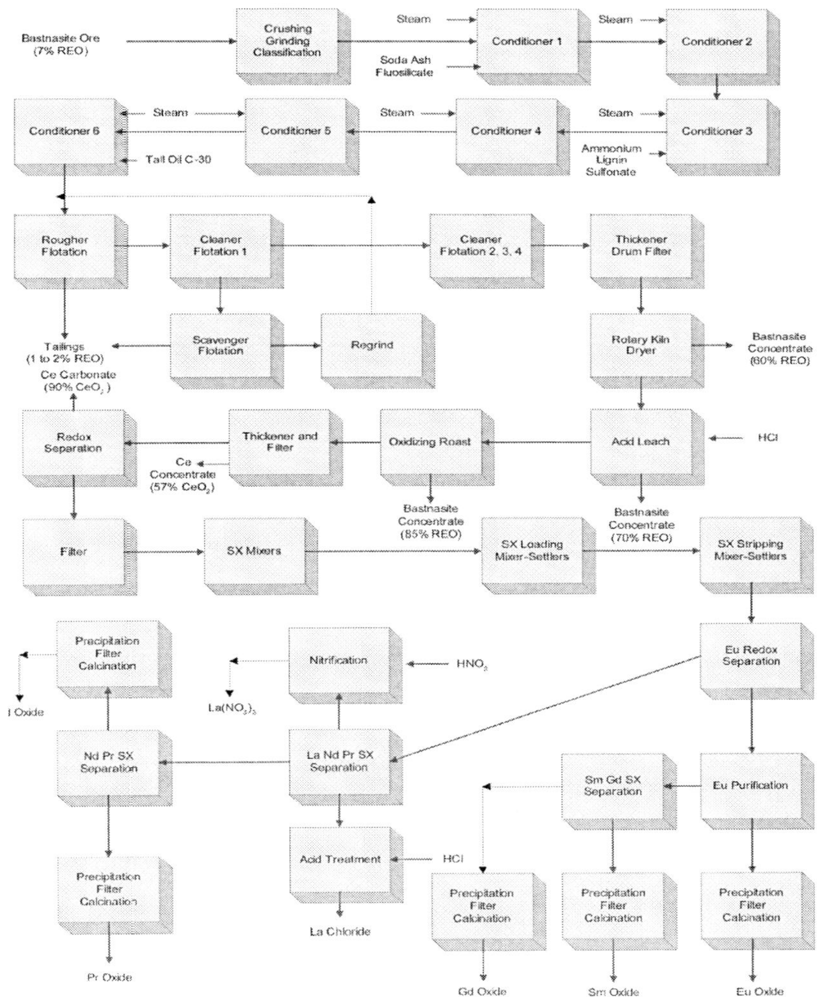

Figure 2. Rare earth elements mineral-processing flow sheet for the Mountain Pass mine, California, about 1995. From one type of ore, no less than 12 rare earth elements products were obtained. REO, rare earth oxides; Ce, cerium; Eu, europium; Gd, gadolinium; La, lanthanum; Nd, neodymium; Pr, praseodymium; Sm, samarium; HCl, hydrochloric acid. (Gupta and Krishnamurthy (2005), Castor and Hedrick (2006)).

Table 3. Rare Earth Elements, Thorium, and Uranium Content of Minerals Found in Rare Earth Elements Deposits

[--, not available; REO, rare earth elements oxides. Minerals in bold have historically been processed to recover rare earth elements. Small quantities of other minerals may be found in deposits that are or have been mined or in unmined deposits]

Mineral	Formula	Mineral Formula		
		REO	ThO_2	UO_2
Allanite (Ce)	$(Ce,Ca,Y)_2(Al,Fe^{2+},Fe^{3+})_3(SiO_4)_3(OH)$	3 to 51	0 to 3	--
Allanite (Y)	$(Y,Ce,Ca)_2(Al,Fe^{3+})_3(SiO_4)_3(OH)$	3 to 51	0 to 3	--
Anatase	$(Ti,REE)O_2$	--	--	--
Ancylite (Ce)	$SrCe(CO_3)_2OH \cdot H_2O$	46 to 53	0 to 0.4	0.1
Bastnasite (Ce)	$(Ce,La)(CO_3)F$	70 to 74	0 to 0.3	0.09
Brannerite	$(U,Ca,Y,Ce)(Ti,Fe)_2O_6$	--	--	--
Britholite (Ce)	$(Ce,Ca)_5(SiO_4,PO_4)_3(OH,F)$	56	1.5	--
Brockite	$(Ca,Th,Ce)(PO_4) \cdot H_2O$	--	--	--
Calcio-ancylite (Ce)	$(Ca,Sr)Ce_3(CO_3)_4(OH)_3 \cdot H_2O$	60	--	--
Cerianite (Ce)	$(Ce^{4+},Th)O_2$	--	--	--
Cerite (Ce)	$Ce_9Fe^{3+}(SiO_2)6[(SiO_3)(OH)](OH)_3$	--	--	--
Cheralite	$(Ca,Ce,Th)(P,Si)O_4$	--	≤30	--
Chevkinite	$(Ca,Ce,Th)_4(Fe^{2+},Mg)_2(Ti,Fe^{3+})3Si_4O_{22}$	--	--	--
Churchite (Y)	$YPO_4 \cdot H_2O$	--	--	--
Crandallite	$CaAl_3(PO_4)_2(OH)_5 \cdot H_2O$	--	--	--
Doverite	$YCaF(CO_3)_2$	--	--	--
Eudialyte	$Na_4(Ca,Ce)_2(Fe^{2+},Mn^{2+},Y)ZrSi_8O_{22}(OH,Cl)_2$	1 to 10	--	--
Euxenite (Y)	$(Y,Ca,Ce,U,Th)(Nb,Ta,Ti)_2O_6$	--	--	--
Fergusonite (Ce)	$(Ce,La,Y)NbO_4$	--	--	--
Fergusonite (Y)	$YNbO_4$	--	--	--
Florencite (Ce)	$CeAl_3(PO_4)_2(OH)_6$	--	1.4	--
Florencite (La)	$(La,Ce)Al_3(PO_4)_2(OH)_6$		1.4	
Fluocerite (Ce)	$(Ce,La)F_3$	--	--	--
Fluocerite (La)	$(La,Ce)F_3$	--	--	--
Fluorapatite	$(Ca,Ce)_5(PO_4)_3F$	0 to 21	0 to 0.01	
Fluorite	$(Ca,REE)F$	--	--	--
Gadolinite (Y)	$Y_2Fe^{2+}Be_2Si_2O_{10}$	40	--	--

Table 3. (Continued

Mineral	Formula	Mineral Formula		
		REO	ThO^2	UO^2
Gagarinite (Y)	$NaCaY(F,Cl)_6$	--	--	--
Gerenite (Y)	$(Ca,Na)_2(Y,REE)_3Si_6O_{18}\cdot 2H_2O$	--	--	--
Gorceixite	$(Ba,REE)A_{13}[(PO_4)_2(OH)_5]\cdot H_2O$	--	--	--
Goyazite	$SrA_{13}(PO_4)_2(OH)_5\cdot H_2O$	--	1.4	--
Hingganite (Y)	$(Y,Yb,Er)_2Be_2Si_2O_8(OH)_2$	--	--	--
Iimoriite (Y)	$Y_2(SiO_4)(CO_3)$	--	--	--
Kainosite (Y)	$Ca_2(Y,Ce)_2Si_4O_{12}(CO_3)\cdot H_2O$	--	--	--
Loparite (Ce)	$(Ce,Na,Ca)(Ti,Nb)O_3$	32 to 34	--	--
Monazite (Ce)	$(Ce,La,Nd,Th)PO_4$	35 to 71	0 to 20	0 to 16
Parisite (Ce)	$Ca(Ce,La)_2(CO_3)_3F_2$	59	0 to 0.5	0 to 0.3
Perovskite	$(Ca,REE)TiO_3$	≤37	0 to 2	0 to 0.05
Pyrochlore	$(Ca,Na,REE)_2Nb_2O_6(OH,F)$	--	--	--
Rhabdophane (Ce)	$(Ce,La)PO_4\cdot H_2O$	--	--	--
Rhabdophane (La)	$(La,Ce)PO_4\cdot H_2O$	--	--	--
Rinkite (rinkolite)	$(Ca,Ce)_4Na(Na,Ca)_2Ti(Si_2O_7)_2F_2(O,F)_2$	--	--	--
Samarskite	$(REE,Fe_2+,Fe_3+,U,Th,Ca)(Nb,Ta,Ti)O_4$	--	--	--
Sphene (titanite)	$(Ca,REE)TiSiO_5$	≤3	--	--
Steenstrupine (Ce)	$Na_{14}Ce_6Mn_2Fe_2(Zr,Th)(Si_6O_{18})_2(PO_4)_7\cdot 3H_2O$			
Synchysite (Ce)	$Ca(Ce,La)(CO_3)_2F$	49 to 52	1.6	--
Synchysite (Y) (doverite)	$Ca(Y,Ce)(CO_3)_2F$	--	--	--
Thalenite (Y)	$Y_3Si_3O_{10}(F,OH)$	--	--	--
Thorite	$(Th,U)SiO_4$	≤3	--	10 to 16
Uraninite	$(U,Th,Ce)O_2$	--	--	--
Vitusite (Ce)	$Na_3(Ce,La,Nd)(PO_4)_2$	--	--	--
Xenotime (Y)	YPO_4	52 to 67	--	0 to 5
Yttrofluorite	$(Ca,Y)F_2$	--	--	--
Zircon	$(Zr,REE)SiO_4$	--	0.1 to 0.8	--

Lateritic deposits—highly weathered soil horizons, rich in iron and aluminum oxide minerals, which develop in a tropical or forested warm environment—have been studied as a potential source of REE; these lateritic REE deposits may contain large resources when they overlie low- grade primary sources, such as carbonatites and syenites. At present, however, only two districts (both in southern China) have been mined in this capacity. These surficial clay deposits account for 14 percent of Chinese REE production (Wu and others, 1996). The ore is referred to as REEbearing ionic absorption clay and forms weathering crusts over granite (Ren, 1985; Wu and others, 1996). Laterite clays from Longnan in the Jiangxi Province yield heavy REE- and Y-rich material whereas ore from Xunwu is light REE–rich (O'Driscoll, 2003).

The relative abundance of rare earth elements within and among deposits is highly variable, but light rare earth elements are typically more abundant than heavy rare earth elements.

Table 4. Dominant Rare Earth Elements–Bearing Minerals Identified in the United States

Mineral	Formula
	Oxides
Aeschynite	$(Ce,Th,Ca...)[(Ti,Nb,Ta)_2O_6]$
Euxenite	$(Y,Er,Ce,U,Pb,Ca)(Nb,Ta,Ti)_2(O,OH)_6$
Fergusonite	$YnbO_4$
Samarskite	$(Y,Er,Fe,Mn,Ca,U,Th,Zr)(Nb,Ta)_2(O,OH)_6$
	Carbonates
Ancylite	$Sr(Ce,La)(CO_3)_2(OH)\cdot(H_2O)$
Bastnasite	$(Ce, La,Y)CO_3F$
Parisite	$Ca(Ce,La)_2(CO_3)_3F_2$
Synchisite	$Ca(Ce,Nd,Y,La)(CO_3)_2F$
Tengerite	$Y_2(CO_3)_3\bullet n(H_2O)$
	Phosphates
Britholite	$(Na,Ce,Ca)_5(OH)[(P,Si)O_4]_3$
Florencite	$(La,Ce)Al_3(PO_4)_2(OH)_6$
Monazite	$(Ce,La,Th,Nd,Y)PO_4$
Xenotime	YPO_4
	Silicates
Allanite	$Ca(Ce,La,Y,Ca)Al_2(Fe^{2+},Fe^{3+})(SiO_4)(Si_2O_7)O(OH)$
Kainosite	$Ca_2(Ce,Y)_2(SiO_4)_3CO_3\cdot H_2O$
Thalenite	$Y2[Si_2O_7]$

The relative proportion of the different rare earth elements in an orebody is quite variable (table 5). The chief differences can be seen in the relative proportion of light to heavy rare earth elements. REE orebodies are typically somewhat enriched in the light REE—lanthanum to gadolinium— compared with average crustal abundances. In comparison, most rare earth ores are notably depleted in the heavy REE— terbium to lutetium. A minority of deposits are relatively enriched in heavy REE, most commonly those that contain xenotime as the principal REE mineral.

Table 5. Distribution of Rare Earth Elements in Selected Rare Earth Elements Deposits

[Rare earth elements listed in order of increasing atomic number; yttrium (Y) is included with these elements because it shares chemical and physical similarities with the lanthanides]

Rare earth element	Average upper crustal Abundance (percent) Wedepohl (1995)	Bastnasite Mountain Pass, USA (percent) Castor (2008)	Bastnasite Bayan Obo, China (percent) Hedrick (2004)	Monazite Green Cove Spring, USA (percent) Hedrick (2004)	Xenotime Lehat, Malaysia (percent) Hedrick (2004)	High Y REE laterite Longnan, China (percent) Hedrick (2004)	Low Y REE laterite Xunwu, China (percent) Hedrick (2004)	Bulk ore Bear Lodge, USA (percent) Castor (2008)	Bulk ore Strange Lake, Canada (percent) Castor (2008)
Lanthanum	19.3	33.8	23.0	17.5	1.2	1.8	43.4	30.4	4.6
Cerium	39.2	49.6	50.0	43.7	3.1	0.4	2.4	45.5	12.0
Praseod-ynium	3.8	4.1	6.2	5.0	0.5	0.7	9.0	4.7	1.4
Neody-mium	15.5	11.2	18.5	17.5	1.6	3.0	31.7	15.8	4.3
Samarium	2.8	0.9	0.8	4.9	1.1	2.8	3.9	1.8	2.1
Europium	0.6	0.1	0.2	0.2	Trace	0.1	0.5	0.4	0.2
Gadoli-nium	1.7	0.2	0.7	6.6	3.5	6.9	3.0	0.7	2.5
Terbium	0.3	0.0	0.1	0.3	0.9	1.3	Trace	0.1	0.3
Dyspro-sium	1.7	0.0	0.1	0.9	8.3	6.7	Trace	0.2	8.2
Holmium	0.4	0.0	Trace	0.1	2.0	1.6	Trace	0.0	1.7
Erbium	1.3	0.0	Trace	Trace	6.4	4.9	Trace	0.0	4.9
Thulium	0.2	0.0	Trace	Trace	1.1	0.7	Trace	<0.01	0.7
Ytterbium	0.9	0.0	Trace	0.1	6.8	2.5	0.3	0.5	4.0
Lutetium	0.2	Trace	Trace	Trace	1.0	0.4	0.1	<0.01	0.4
Yttrium	12.3	0.1	Trace	2.5	61.0	65.0	8.0	<0.01	52.8

Given that each individual rare earth element has its own particular uses and market (table 6), the proportions of the various rare earth elements in a deposit are unlikely to parallel those of demand for rare earth elements. For instance, the most abundant rare earth element, cerium, is available in quantities that exceed demand for traditional uses (Heymann, 2010). Most REE deposits currently (2010) considered for development are enriched in light REE and would likely flood the market for cerium if put into production. By contrast, heavy REE are in short supply with limited reserves. Certain rare earth elements, such as lutetium, presently have no market and are not worth recovering at this time.

Rare earth elements are typically obtained as a byproduct or coproduct of mining other mineral commodities.

Table 6. Useage of Rare Earth Elements

[Each rare earth element has its own applications and market.

Application	La	Ce	Pr	Nd	Sm	Eu	Gd	Tb	Dy	Y	Other
Magnets	--	--	23.4	69.4	--	--	2	0.2	5	--	--
Battery alloys	50	33.4	3.3	10	3.3	--	--	--	--	--	--
Metal alloys	26	52	5.5	16.5	--	--	--	--	--	--	--
Auto catalysts	5	90	2	3	--	--	--	--	--	--	--
Petroleum refining	90	10	--	--	--	--	--	--	--	--	--
Polishing compounds	31.5	65	3.5	--	--	--	--	--	--	--	--
Glass additives	24	66	1	3	--	--	--	--	--	2	4
Phosphors	8.5	11	--	--	--	4.9	1.8	4.6	--	69.2	--
Ceramics	17	12	6	12	--	--	--	--	--	53	--
Other	19	39	4	15	2	--	1	--	--	19	--

Source: Lynas Corporation (2010)]

Table 7. Production of Rare Earth Elements Mines in 2009

Country	Mine	2009 output (metric tons TREO)	Primary product	Byproduct
Brazil	Buena Norte	650	Ilmenite concentrate	Monazite concentrate.
China	Bayan Obo	55,000	Iron ore	Bastnäsite concentrate.
	Sichuan[1]	10,000	Bastnäsite concentrate	
	South China[1]	45,000	Rare earth elements	
India	Heavy-mineral sands	2,700	Ilmenite concentrate	Monazite concentrate.
Malaysia	Ipoh sand plant	380	Cassiterite concentrate	Xenotime concentrate.
Russia	Lovozero	2,500	Loparite concentrate	Rare earth elements chloride.

[1]Many small producers and a few medium-large producers. The Chinese rare earth elements–mining industry is currently (2010) undergoing government- directed rationalization to reduce the number of producers.

Source: USGS Mineral Commodity Summaries (U.S. Geological Survey, 2010). TREO, total rare earth elements oxide.

When the economic viability of a mining project is assessed, the potential mineral products are divided into principal products and byproducts. The principal product, for example zinc in a zinc mine, contributes most to the value of the minerals produced. Generally, returns from the principal product are sufficient to pay the costs of mining and processing. All other products are referred to as byproducts, whose returns typically bolster the overall profitability of a mine. Where two or more products of essential value are obtained, they are called coproducts. A salient feature of rare earth elements mining is that REE-rich minerals may be byproducts or coproducts of mining other mineral commodities.

Mine production decisions are driven by demand for principal products, not for byproducts. Thus, production of byproduct REE will vary subject to changes in demand for principal products and will be relatively unresponsive to demand for REE. China currently (2010) accounts for about 96 percent of global rare earth elements production (table 7). Of a total production of 120,000 metric tons, about 55,000 metric tons was produced as a byproduct of the Bayan Obo iron mine. This fact means that at least 44 percent of world rare earth elements production is a by-product. Of the remaining Chinese production, about 25,000 metric tons is produced in southern China as a primary product from ion-adsorption deposits. The status of remaining Chinese production is unclear. The balance of global REE production is as a byproduct. Conceivably, as much as 90 percent of global rare earth elements production is as a byproduct or coproduct.

CURRENT SOURCES AND DOMESTIC RESERVES

The United States currently imports all of its rare earth elements (REE) raw materials from foreign sources, principally China (U.S. Geological Survey, 2010). This has not always been the case. The USGS annually reports global and domestic production and trade in REE in its publications Minerals Yearbook and Mineral Commodity Summaries. Prior to 1998, when production from the Mountain Pass mine in California was curtailed, the United States produced most of the light REE consumed domestically and by free market countries. Heavy REE were obtained from imported monazite concentrates. That changed in the 1980s after China became the dominant global supplier of light and heavy REE (Papp and others, 2008). In 2002, the Mountain Pass mine in California, the sole domestic producer of REE minerals, shut down. Although the mine has continued to produce REE materials from stockpiled raw materials, no new REE ores are being mined. Since then, the United States has obtained all of its REE raw materials from imports, principally from China. China accounts for 95 percent of global REE production despite having only 36 percent of identified world reserves (table 8).

Concentration of Supply

The high concentration of production of REE in one country is not unusual for a minor metal commodity. For example, a single mine in the United States supplies 86 percent of world demand for beryllium and two mines in Brazil account for 92 percent of world niobium production (U.S. Geological Survey, 2010). Such concentration of supply, which has long been of concern in regard to price manipulation, also raises issues related to reliability of supply. Given an equal risk of a natural disaster, industrial accident, labor strike, political strife, or anything else that might interrupt production, a single source of supply is inherently more risky than multiple sources of supply. Even though these various risks are not equal among countries, concentration of supply is a key indicator of mineral-supply risk.

Table 9 compares the supply situation of REE with other internationally traded minerals using several measures of concentration. These measures are used by economists to study market concentration and by regulators for antitrust purposes. In table 9, concentration ratios, abbreviated CR2 and CR3, measure the total percent share in United States imports and world production of the top two or top three supplier countries, respectively. A high percentage, such as the CR2 of 94 percent and CR3 of 96 percent shown for REE (excluding yttrium and scandium), indicates that imports and world production are principally derived from one or two countries. A third measure is the Herfindahl index (Stigler, 1983), which was originally developed to measure the degree of competition in an industry. It is calculated according to the equation

$$HI = \sum_{i=1}^{n} s_i^2$$

where s_i is the share in global production or United States imports by country i with n countries. The larger this index, the more concentrated are world production and United States imports by country. The Herfindahl index can be normalized

$$NHI = \frac{HI - \frac{1}{n}}{1 - \frac{1}{n}}$$

such that the index ranges from 0 to 1.0, which facilitates comparison between different mineral commodities. A normalized Herfindahl Index of 1.0 indicates concentration in a single country; an index of 0 indicates that all countries have exactly the same share in United States imports or world production.

As shown in table 9, all three of these indices place REE (including yttrium) at the top of all mineral commodities in terms of concentration of United States imports and world production. Antimony and niobium, which are mostly produced in China and Brazil, respectively, have very similar concentration indices. Rhenium is an example of a mineral commodity that the

United States largely imports from a single country, Chile, but whose global production is not particularly concentrated.

Table 8. World Production and Reserves of Rare Earth Elements Minerals in 2009

[In 2009, China produced 95 percent of world rare earth elements although it had only 36 percent of rare earth elements reserves. TREO, total metric tons of rare earth oxides]

Country	2009			
	Production		Reserves	
	TREO (metric tons)	Share (percent)	TREO (metric tons)	Share (percent)
Australia	0	0	5,400,000	5
Brazil	650	0.5	48,000	0.05
China	120,000	95	36,000,000	36
Commonweath of Independent States	2,500	2	19,000,000	19
India	2,700	2	3,100,000	3
Malaysia	380	0.3	30,000	0.03
United States	0	0	13,000,000	13
Other	0	0	22,000,000	22
Total	126,230		99,000,000	

On the basis of these data, it is no exaggeration to say that China dominates the world REE industry. This dominance is attributable to China's large, high-quality resources of REE coupled with minimal capital investment, low labor costs, and lack of environmental regulation (Hurst, 2010). Referring back to table 8, China has only about a third of global REE; hence, a lower cost of production is a reasonable explanation for China's position as the world's dominant REE producer. Papp and others (2008) show that REE prices dropped dramatically from 1997 to 2008, consistent with the introduction of significant amounts of lower priced Chinese REE.

Table 9. Measures of Concentration for Selected World Metal Mining Industries

[CR2 and CR3, two-county and three-country concentration ratios, respectively. NHI, normalized Herfindahl index. The higher the index, the more concentrated are mineral production and United States imports. CR2 and CR3 are rounded to the nearest percent resulting, in some cases, in a slight discrepancy between the concentration ratios and the normalized Herfindahl index. RI, country risk index. See text for an explanation of indices. Data are for 2007, the latest year for which complete information is available from the U.S. Geological Survey (2010)]

Mineral commodity	Import reliance (percent)	United States imports (percent)				World production (percent)			
		CR2	CR3	NHI	RI	CR2	CR3	NHI	RI
Antimony	86	90	98	0.42	1.9	91	94	0.77	2.3
Bauxite and alumina	100	50	64	0.19	4.6	46	58	0.16	2.8
Bismuth	95	62	80	0.26	0.8	75	90	0.29	2.3
Cobalt	78	43	56	0.13	1.7	52	63	0.20	1.4
Copper	37	75	88	0.32	1.5	44	51	0.16	2.4
Gallium	99	57	73	0.21	1.3	51	65	0.19	1.9
Indium	100	72	81	0.31	1.3	68	76	0.36	1.4
Manganese	100	54	65	0.21	2.9	46	64	0.17	2.8
Nickel	17	59	68	0.23	1.0	32	46	0.10	2.6
Niobium	100	96	97	0.79	2.7	100	100	0.90	2.9
Platinum	94	50	65	0.17	1.5	91	94	0.63	2.9
Rare earth elements	100	94	96	0.83	1.9	99	100	0.94	2.0
Rhenium	86	95	98	0.81	1.8	59	68	0.26	2.3
Tantalum	100	35	50	0.13	1.6	75	85	0.35	2.0
Tin	79	69	79	0.31	3.2	74	91	0.30	3.3
Titanium	64	85	94	0.39	3.3	55	77	0.23	2.2
Tungsten	70	50	69	0.19	3.3	81	86	0.57	2.3
Vanadium	100	66	74	0.35	1.5	72	97	0.33	1.5
Yttrium	100	96	99	0.78	1.8	100	100	0.98	2.0
Zinc	58	67	82	0.19	1.1	52	66	0.19	1.9

Risk of Supply Interruption

Assessing our nation's vulnerability to mineral-supply disruptions is a classic exercise in risk analysis. The analysis has two components: the nature and probability of threats, and assessment of potential impact. Quantitative measurement of these components would be useful in ranking the relative security of supply of the various mineral commodities used and imported by the United States. A first step was made by a special committee of the National Research Council when it recommended the criticality matrix as a tool for assessing mineral supply risk (National Research Council, 2008). The criticality matrix is a plot that subjectively contrasts supply risk on one axis and the effect of supply restriction on the other. The authors of the study ranked various imported metals, including REE, on a scale of 1 to 4 (low to high) as shown in figure 3. Rare earth elements were ranked 4 (high) for supply risk and 3 (moderately high) for effect of supply restriction.

An analysis of the effect of supply restrictions requires a level of economic analysis that is outside the traditional responsibilities of the U.S. Geological Survey and beyond the scope of this report. Long (2009) proposed combining the quantitative measures of concentration discussed above with measures of country risk to obtain a relative ranking of minerals by supply risk. A similar approach was independently adopted by the Raw Materials Supply Group of the European Union in a recent study of European mineral security (Raw Materials Supply Group, 2010). The European Union study does include a rough measure of the economic effect of a mineral supply disruption.

There are many measures of country risk from which to choose. Long (2009) used the Country Risk Classification that is published annually by the Organization for Economic Co-operation and Development (Organization for Economic Co-Operation and Development, 2008). This classification is a measure of a country's credit risk or likelihood that a country will service its external debt. Countries are subjectively ranked on a scale from 0 to 7, where 0 is the lowest degree of risk and 7 the highest. An aggregate country risk index for a commodity is obtained as the sum of individual country risk indices weighted by share in United States imports or world production (table 9). This aggregate country risk index likewise falls on the scale of 0 to 7. Table 9 illustrates how these indices and ratios can be used. Comparison of concentration indices for United States imports and for world production shows whether imports are more concentrated than global production, thus indicating opportunities for further diversification of supply. A high import concentration index with a low country risk index suggests that imports are obtained from stable trading partners such as Canada and Australia. High indices across the board are cause for greatest concern and indicate those commodities that are of greatest risk.

The European Union study (Raw Materials Supply Group, 2010) used the World Bank's World Governance Indicators as a measure of political risk. The World Bank estimates six governance indicators: voice and accountability, political stability, government effectiveness, regulatory quality, rule of law, and control of corruption (World Bank, 2010). The European Union study unfortunately does not specify which indicator was used or, if all indicators were used, how they were combined. The U.S. Geological Survey has identified other indicators of country risk, such as the Economic Freedom Score (Heritage Foundation, 2010), the Corruption Perceptions Index (Transparency International, 2010), and rankings of countries for mining investment (Behre Dolbear, 2010; McMahon and Cervantes, 2010).

Aside from an indicator of country risk, the European Union study also used a subjective ranking of the degree to which other minerals can substitute for the mineral in question, measures of recycling rates and environmental policy risk, and a rescaled Herfindahl index to measure mineral supply concentration. Using these indicators, some 14 metals and minerals were shortlisted as critical raw materials for European Union member nations, "critical" in this case signifying a high degree of both supply risk and economic importance. The shortlisted metals and minerals are antimony, beryllium, cobalt, fluorspar, gallium, germanium, graphite, indium, magnesium, niobium, platinum group metals, REE, tantalum, and tungsten (Raw Materials Supply Group, 2010). Long (2009) listed 15 metals and minerals: antimony, barite, chromite, cobalt, fluorspar, gallium, graphite, indium, niobium, platinum group metals, REE, rhenium, tantalum, titanium, and tungsten as those minerals with the greatest supply risk to the United States economy. The two lists are very similar. The differences are beryllium, which is imported by Europe from the United States; barite, critical to the US oil and gas industry; and chromite, rhenium, and titanium, whose supply was rated as less risky in the European Union report.

In both the USGS and European Union studies of mineral supply risk, REE rank highest as mineral raw materials of critical concern, given uncertain future supplies and their importance to advanced industrial economies. Neither of these studies addressed measures to mitigate these risks but each did recommend further study, including examination of mineral policy options. There are geologic factors, however, that should be considered in future studies of REE supply, such as the extent and quality of domestic REE resources, undeveloped resources in other low-risk countries, and the time it takes to develop these resources into producing mines.

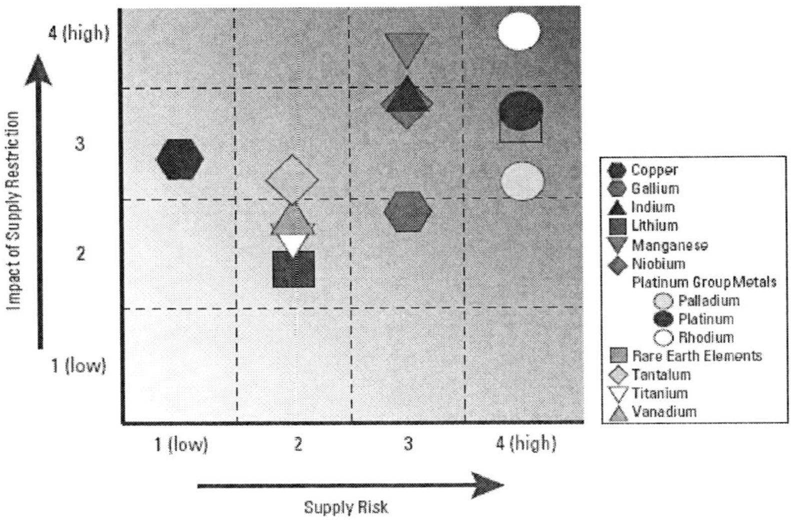

Figure 3. Criticality matrix for selected imported metals (National Research Council, 2008).

Domestic and World Resources

The main body of this report is a review of the geology and known mineral resources of the principal domestic United States deposits of REE minerals discovered to date (2010). These resources are summarized in table 10. It is important to recognize that resource estimates are of differing accuracy and reliability, depending on the degree of exploration undertaken to date. Many of the estimates in table 10 are obtained by inference from surface exposures of mineralization, a small number of samples, and inferences or assumptions about how deep mineralization extends. Some deposits have been explored at depth by core drilling. A very few have been drilled on a narrowly spaced grid sufficient for an estimate of how much mineralized material may be economic to mine. The potential economic viability of any of these resources can be reliably assessed only with sufficient drilling, pilot plant–scale metallurgical testing, and definitive economic analysis. Only one domestic deposit, Mountain Pass, California, meets those criteria and can be reported to contain a sizable reserve of REE-bearing ore.

Table 11 reports reserves and resources in REE deposits worldwide, divided into three categories. The first category comprises deposits sufficiently explored to estimate a mine plan resource. Although a mine has been designed or already exists for each of these deposits, they have not been demonstrated to be economically viable by means of a definitive feasibility study; hence, they are classed as resources. The second category comprises measured, indicated, and inferred resources for well-explored deposits that have not yet been subject to a feasibility study that includes a mine design. The third category, unclassified resources, is a mixed bag of known resources that are unlikely to be exploited, such as Pilanesberg, South Africa, which is now within a national park, and the Olympic Dam mine, Australia, where extensive study has found that REE are not economic to recover even as a byproduct. Other deposits in this category have been little explored and the resources are inferred from surface exposures and limited sampling. No reliable data are available for mines and deposits in China, Russia, and North Korea.

The first two categories of resources are the only short- and medium-term sources of additional REE that might contribute to the global supply. These categories will likely be augmented through further exploration at existing mines and development projects. It is possible that long-term supply can be met through exploration of known deposits that have had little or no drilling and by the discovery of new deposits. The projects listed in the first two of the categories shown in tables 10 and 11 put an upper limit on a near-term potential for production of REE mineral supplies. This limit can be put at about 14 million tons of contained total REE oxides (TREO), with a country risk index of near zero because almost all of that production potential is in the United States, Australia, and Canada.

Long-term prospects for the discovery of new reserves and resources depend on sufficient exploration. About 150 projects worldwide that are prospecting and exploring for rare earth elements are known to the U.S. Geological Survey (U.S. Geological Survey, 2010). Most of this activity began in the last 2 years and it will be some time before results are known. Whether this exploration further diversifies REE mineral supplies depends on discovering deposits of adequate size and quality in the right countries. Probability of discovery requires a quantitative mineral resource assessment, which has never been done for REE minerals in any country. The last REE deposit discovered and developed into a mine in the United States was the Mountain Pass mine in California, discovered in 1949 and put into production in 1953. That was more than 50 years ago and is not indicative of the time required to discover and develop REE deposits in today's regulatory climate. During the past 50 years outside of China, there has been little REE exploration and almost no mine development; hence, we have no real REE exploration and development record to draw upon for assessing the future pace of discovery and development.

Table 10. Domestic Reserves and Resources of Rare Earth Elements, Excluding Heavy-Mineral Placer and Phosphate Deposits

[TREO, total rare earth oxides. Reserves proven and probable classified according to definitions and standards of the Securities and Exchange Commission (Securities and Exchange Commission = http://www.sec.gov/index.htm). Inferred resources classified according to the standards of Canadian National Instrument 43-101 (Canadian National Instrument 43-101 = http://www.ccpg.ca/profprac/index.php?lang=en&subpg=natguidelines). Unclassified resources based on little or no drilling. For data on resources in heavy-mineral placer and phosphate deposits, which are not of economic interest, see Jackson and Christiansen (1993)]

Deposit		Tonnage (metric tons)	Grade (percent TREO)	Contained TREO (metric tons)	Source
Reserves—Proven and probable					
Mountain Pass,	California	13,588,000	8.24	1,120,000	Molycorp, Inc. (2010).
Resources—Inferred					
Bear Lodge,	Wyoming	10,678,000	3.60	384,000	Noble and others (2009).
Resources—Unclassified					
Bald Mountain,	Wyoming	18,000,000	0.08	14,400	Osterwald and others (1966).
Bokan Mountain,	Alaska	34,100,000	0.48	164,000	Keyser and Kennedy (2007).
Diamond Creek,	Idaho	5,800,000	1.22	70,800	Staatz and others (1979).
Elk Creek,	Nebraska	39,400,000			Molycorp, Inc. (1986).
Gallinas Mtns.,	New Mexico	46,000	2.95	1,400	Jackson and Christiansen (1993).
Hall Mountain,	Idaho	100,000	0.05	50	Staatz and others (1979).
Hick's Dome,	Illinois	14,700,000	0.42	62,000	Jackson and Christiansen (1993).
Iron Hill,	Colorado	2,424,000,000	0.40	9,696,000	Staatz and others (1979).
Lemhi Pass,	Idaho	500,000	0.33	1,650	Staatz and others (1979).
Mineville,	New York	9,000,000	0.9	80,000	McKeown and Klemic (1956).
Music Valley,	California	50,000	8.6	4,300	Jackson and Christiansen (1993).
Pajarito,	New Mexico	2,400,000	0.18	4,000	Jackson and Christiansen (1993).
Pea Ridge,	Missouri	600,000	12	72,000	Grauch and others (2010).
Scrub Oaks,	New Jersey	10,000,000	0.38	38,000	Klemic and other (1959).
Wet Mou-ntains,	Colorado	13,957,000	0.42	59,000	Jackson and Christiansen (1993).

Table 11. Reserves and Resources of Rare Earth Elements Outside of the United States, Excluding Heavy-Mineral Placer and Phosphate Deposits

[TREO, total rare earth elements oxides. Heavy-mineral placers are mined for rare earth elements in only a few places, such as India and Malaysia, and reserve information is unavailable. Reserves and resources classified according to one of several national standards, such as Canadian National Instrument 43-101, JORC, and SAMREC codes. (JORC, The Australasian Code for Reporting of Exploration Results, Mineral Resources and Ore Reserves (JORC Code, 2004 edition) (http://www.jorc.org/); SAMREC, South African Mineral Resource Committee, http://www.samcode.co.za/downloads/SAMREC2009.pdf).) Unclassified resources based on little or no drilling. Reliable data on rare earth elements reserves and resources in China, North Korea, and Russia are not available. Where reserves and resources are given for the same deposit, resources include reserves. For data on other resources in heavy-mineral placers and in phosphate deposits, see Jackson and Christiansen (1993)]

Deposit		Tonnage (metric tons)	Grade (percent TREO)	Contained TREO (metric tons)	Source
Measured-in-pit resources					
Brockman,	Australia	4,290,000	0.2	8,600	Chalmers (1990).
Mount Weld,	Australia	2,100,000	15.5	326,000	Lynas Corporation (2010).
Thor Lake (Lake Zone),	Canada	12,010,000	1.70	204,000	Paul and Stubens (2009).
Steenkampskraal,	South Africa	249,500	11.80	29,500	Great Western Minerals Group Ltd. (2009).
Measured, indicated, inferred resources					
Brockman,	Australia	50,000,000	0.23	115,000	Chalmers (1990).
Cummins Range,	Australia	4,170,000	1.72	72,000	Navigator Resources Ltd.(2009).
Dubbo,	Australia	73,200,000	0.89	651,500	Alkane Reso-urces (2010).

Table 11. (Continued)

Deposit		Tonnage (metric tons)	Grade (percent TREO)	Contained TREO (metric tons)	Source
Mount Weld,	Australia	15,020,000	8.60	1,292,000	Lynas Corporation (2010).
Narraburra,	Australia	55,000,000	0.03	16,500	Capital Mining Ltd. (2009).
Nolans Bore,	Australia	30,300,000	2.80	849,000	Arafura Resources Ltd. (2010).
Hoidas Lake,	Canada	2,847,000	2.00	57,000	Dunn (2009).
Strange Lake,	Canada	137,639,000	0.97	1,335,000	Daigle and Maunula (2010).
Thor Lake (Lake Zone),	Canada	175,930,000	1.43	2,516,000	Paul and Stu-bens (2009).
Thor Lake (North T),	Canada	1,136,000	0.71	8,000	Palmer and Broad (2007).
Zeus (Kipawa),	Canada	2,270,000	0.11	2,500	Knox and others (2009).
Kvanefjeld,	Greenland	457,000,000	1.07	4,890,000	Greenland Minerals and Energy Ltd. (2009).
Kangankunde Hill,	Malawi	2,500,000	4.24	107,000	Lynas Corpo-ration Ltd. (2007).
Unclassified resources					
John Galt,	Australia	382,000	7.96	30,400	Northern Uranium Ltd. (2010).
Olympic Dam,	Australia	>2,000,000,00 0	0.50	>10,000,000	Oreskes and Einaudi (1990).
Yangibana,	Australia	3,500,000	1.70	59,500	Jackson and Christiansen (1993).
Araxá,	Brazil	450,000,000	1.80	8,100,000	Filho and others (2005).
Catalão I,	Brazil	10,000,000	0.90	90,000	Hirano and others (1990).
Pitinga,	Brazil	164,000,000	0.15	246,000	Bastos Neto and Pereira (2009).

Table 11. (Continued)

Deposit		Tonnage (metric tons)	Grade (percent TREO)	Contained TREO (metric tons)	Source
Poços de Caldas,	Brazil			115,000	Wedow (1967).
Seis Lagos,	Brazil	2,900,000,000	1.50	43,500,000	De Sousa (1996).
Tapira,	Brazil	5,200,000	10.5	546,000	Hirano and others (1990).
Kasagwe,	Burundi	67,000	1.50	1,000	Jackson and Christiansen (1993).
Oka,	Canada	210,000,000	0.127	267,000	Orris and Grauch (2002).
Mrima Hill,	Kenya	6,000,000	16.2	972,000	Pell (1996).
Ak-Tyuz,	Kyrgyzstan	15,000,000	1.00	150,000	Malyukova and others (2005).
Karajilga,	Kyrgyzstan	957,000	0.70	6,700	Bogdetsky and others (2001).
Kutessai II,	Kyrgyzstan	20,228,000	0.22-0.3	<60,000	Stans Energy Corp. (2010).
Sarysai,	Kyrgyzstan	7,000,000	0.20	14,000	Bogdetsky and others (2001).
Pilanesberg,	South Africa	13,500,000	0.70	94,500	Lurie (1986).
Zandkopsdrift,	South Africa	31,500,000	3.60	1,130,000	Frontier Minerals (2009).
Kizilcaören,	Turkey	4,695,000	2.78	130,500	Morteani and Satir (1989).
Dong Pao,	Vietnam	500,000,000	1.40	7,000,000	Kušnir (2000).
Mau Xe North,	Vietnam	557,000,000	1.40	7,800,000	Kušnir (2000).

Developing Rare Earth Elements Resources

Rare earth elements resources are distributed between many mineral deposits, but only a proportion will be economic to develop and mine. By convention, that portion of resources that is economic to mine is classified as a “reserve.” That a rare earth deposit contains reserves does not mean that it will be developed and mined—it means only that it is economic to do so. Among the many rare earth reserves available, mining companies will select the most profitable to develop, potentially leaving less profitable reserves undeveloped. Reserves may also be undeveloped because of adverse land use restrictions, civil strife, and a host of other political and social factors.

Developing a new mine requires a prolonged effort of prospecting, exploration, process development, feasibility studies, permitting, construction, and commissioning. These efforts are broadly sequential but commonly overlap. The time required to complete all steps is variable but appreciable, particularly compared with the time typically required by non- extractive industries. Studies of the time required to complete this process typically examine the interval between recognition of a potentially economic deposit and the commencement of commercial production. Peters (1966) examined the development history of a large number of mines and divided them into four classes. The first class requires a preproduction period of 2 years or less. This class of mine is characterized by simple ores of high unit value, the applicability of conventional mining and mineral processing methods, the absence of need for much additional transportation or power infrastructure, an assured market, and the requirement for only short-term financing. A modern example in this class would be a small- to medium- sized open pit-heap leach gold deposit in Nevada. Gold has a high unit value and ready market. The pertinent mining and mineral processing technology is also simple and well-known, with excellent local infrastructure, and regulatory authorities have much experience with this type of mining.

The other three categories are characterized by increasing complexity of ores, size of operations, infrastructure and financing needs, and lower unit values. The last category has a preproduction period of 7 years or more. A modern example would be a nickel laterite mine, the metallurgy of which is quite troublesome; many such nickel mines have required in excess of 10 years of process development plus delays because of market timing. Developing a new nickel laterite mine is best timed to open during a period of high nickel prices. A new REE mine would almost certainly fall into this last category for many of the same reasons—complex metallurgy and restricted opportunities for market entry.

The time to develop a mine in each category is likely to take longer today than it did when Peters (1966) did his study. Most modern mines are developed under a comprehensive regulatory environment where, in the United States at least, environmental studies, due diligence studies by financing sources, permitting, public participation, and due process require substantial amounts of time. Further delays may be caused by public controversy and litigation against a proposed mine. These delays are illustrated in table 12, which tallies the time it took to develop metal mines that opened in the United States since 2000. The time to obtain a permit has required as many as 17 years, and one mine, the Pogo, Alaska gold mine, was developed under an expedited permitting schedule that still took 7 years. For a small gold mine in Nevada, once permits were obtained, the time to construct and put a mine into operation took 1 month. For the Kensington, Alaska, gold mine, however, whose operating permits were contested in the courts, the process lasted 63 months. Ramp-up times for new mines took from 2 to 12 months; the longest was for Pogo, Alaska, which had unexpected metallurgical difficulties.

Developing a Rare Earth Elements Mine

The first step in developing a new REE mine is to locate a suitable REE deposit. Exploration for new deposits may be in either of two contexts: greenfield exploration to find new deposits in areas not previously mined and brownfield exploration in and adjacent to current and past mining activity.

Greenfield exploration may target frontier areas with no previous exploration or may follow up on past exploration results. Brownfield exploration includes searching for extensions to known reserves and resources within or near a mine as well as new deposits in the vicinity of existing operations. The latter may extend the life of an existing mine or result in the complete redevelopment of a former mine.

Table 12. Time Required to Obtain Permits, Construct, and Commission Recently Opened Metal Mines in the United States

[NYA, not yet achieved, production not yet begun or commercial operations not achieved; PGE, platinum group elements. Yes, long permitting and development delays because of litigation by government agencies and nongovernmental organizations]

Mine		Commodity	Permitting began	Permitting completed	Production began	Commercial operations began	Litigation reported
Alta Mesa,	Texas	U	1999	2004	10/2005	1/2006	
Arizona 1,	Arizona	U	mid-2007	2009	NYA	NYA	Yes.
Ashdown,	Nevada	Mo Au	2/2004	11/2006	12/2006	NYA	
Buckhorn,	Washington	Au	1992	9/2006	10/2008	11/2008	Yes.
Carlota,	Arizona	Cu	2/1992	6/2007	12/2008	1/2009	Yes.
Eagle,	Michigan	Ni Cu Co PGE	4/2004	1/20 10	NYA	NYA	Yes.
East Boulder,	Montana	PGE	1995	1998	6/200 1	1/2002	
Kensington,	Alaska	Au	3/1988	6/2005	9/2010	NYA	Yes.
Leeville,	Nevada	Au	7/1997	8/2002	10/2006	4th quarter 2006	
Lisbon Valley,	Utah	Cu	2/1996	7/2004	1st quarter 2006	NYA	Yes.
Pend Oreille,	Washington	Zn	1992	9/2000	1/2004	8/2004	
Phoenix,	Nevada	Au	1/1999	1/2004	10/2006	4th quarter 2006	
Pogo,	Alaska	Au	12/1997	4/2004	2/2006	4/2007	
Rock Creek,	Alaska	Au	2003	8/2006	9/2008	NYA	Yes.
Rossi (Storm),	Nevada	Au	1990	3Q/2006	3/2007	12/2007	
Safford,	Arizona	Cu	4/1998	7/2006	4th quarter 2007	2nd half 2008	Yes.
Turquoise Ridge,	Nevada	Au	9/1995	5/2003	2004	NYA	

Exploration is an uncertain process conducted with limited capital. An explorer's objective is to find a deposit of the targeted type, size, and quality with the least amount of expense. Thus, any particular location of merit is rarely exhaustively explored. Exploration will proceed on the basis of favorable indicators so long as objectives are met within budget. An exploration project will be curtailed if evidence is found that contraindicates the prospectivity of the target or if results are insufficient to justify further work. Exploration commonly runs in cycles, prompted by short- to medium-term increases in mineral prices. The low side of a price cycle may prompt little or no exploration activity. Any particular prospect may undergo many episodes of exploration by different parties during many decades; the exploration episodes are motivated by price cycles, new cost-saving technology, improved exploration concepts and methods, and the vagaries of land access and exploration management.

In the case of REE, very little exploration has ever been done and the most intense period ever of direct exploration for REE is currently underway (2010). Many important past discoveries have been serendipitous—the world's largest REE mine, Bayan Obo in China, was first mined for iron ore. The "funny" steel produced from this ore was investigated and found to be contaminated with REE (Laznicka, 2006). Mountain Pass, California, was found in the course of a U.S. Geological Survey radioactivity reconnaissance project that expected to find uranium. Many REE-bearing carbonatite deposits were initially explored as a source of niobium or phosphates.

The discovery of a REE deposit must be proved by exten sive trenching, drilling, and sampling. Drilling will initially be on a widely spaced grid to evaluate the extent and richness of mineralization. If results are favorable, the deposit will be drilled on progressively narrower spacing until a measured resource of adequate size is established. Concurrently, baseline environmental studies will be completed and bulk samples obtained for metallurgical testing. A conceptual mine plan and metallurgical plan will be engineered and a preliminary economic feasibility study undertaken. If the results of these studies are favorable, further work will develop an optimized mine plan, validate metallurgical processing on a pilot plant scale, begin application for permits, and conduct a definitive economic feasibility study. If the results of the feasibility study are positive, financing will be sought and permitting further advanced along with detailed plant engineering. Permitting will typically require an approved plan of operations, a positive environmental impact study, and some kind of final permission by a government agency. If external financing is required, an independent due diligence study will verify the results of the feasibility study.

Once financing and regulatory approval are in place, a construction decision will be announced. Ongoing detailed engineering design will be completed, long lead-time items will be ordered, construction work will be bid on, and contracts will be awarded. Construction begins as soon as possible and is executed quickly to minimize interest and opportunity costs. As mine components are completed, they will be tested and placed in operation with a ramp-up period until full commercial production at the planned output rate is achieved. The ramp-up process may be completed smoothly in a matter of months or may encounter unanticipated difficulties that require an extended period of time to solve. Some mines have failed at this stage because of ore grades that are lower than expected or metallurgical processes that do not work as planned.

Up to this point, a mining company has been spending large amounts of money without any revenue from mineral sales. Capital must be supplied internally, from net revenues of a mining company's other operations, or externally from banks or investors. The cost of borrowing these funds is considerable and repayment cannot even start until a mine goes into production. The largest of currently (2010) proposed new REE mining operations, including Mountain Pass, California, have reported premining capital requirements of a half a billion dollars or more.

A mine will operate until reserves are exhausted. Today, mine closure and reclamation are planned before a mine is built; reclamation is undertaken concurrent with operations wherever possible. Mobile equipment and structures are removed. Roads, waste stockpiles, tailings, and surface facilities are also reclaimed in an attempt to restore the land to its previous use and appearance. Large open pits are not typically backfilled because of the large energy and CO_2 emissions costs of doing so. Instead, to minimize environmental impact they are reclaimed as landscape features.

Only one REE mine, at Mountain Pass, California, has ever been developed in the United States. That occurred in the early 1950s and is not illustrative of what is required to develop such a mine today. Nickel laterite deposits were suggested above as a useful analog for the development of a new REE mine. This comparison is particularly apt for a REE laterite deposit, such as Mount Weld, Australia. Niobium-bearing carbonatites are another close analog. These two deposit types share complex and difficult metallurgy and typically require extended periods of time to develop. The time that was required to bring into production recently developed nickel laterite mines and most niobium carbonatite mines is shown in table 13. All of these mines were developed outside of the United States and the permitting delays noted in table 12 do not pertain in this case.

The time from discovery to initial production for the mines listed in table 13 range from 5 to 50 years, permitting to initial production 1 to 7 years, and ramp-up times were 3 to 42 months, except in the case of Araxá, Brazil, which suffered some unusual political delays. These development times are in line with the results of Peters (1966) and indicate that in some instances a relatively rapid pace of development can be obtained. It should be noted, however, that the most advanced REE projects other than Mountain Pass, California, namely Mount Weld and Dubbo in Australia, have required years of metallurgical testing and development and fall into the fourth, 7-year-or-more category of Peters (1966).

Summary

United States domestic reserves and inferred resources of REE are approximately 1.5 million tons, which are large compared with peak domestic consumption of REE of 10,200 tons in 2007 (U.S. Geological Survey, 2010). How much of that reserve and resource will be economically available, when, and at what rate, cannot be addressed with the data at hand. It can be said that the reserves and inferred resources reported in table 10 are of light REE and that these two potential mines may not be able to meet domestic needs for heavy REE with the production plans currently (2010) proposed. The pipeline of new REE projects within the United States is rather thin, with 10 out of 150 REE exploration projects identified worldwide. If we extend our analysis to reliable trading partners, such as Australia and Canada, prospects for diversifying supply and meeting future demand are considerably improved. Unfortunately, the time required for development of new REE mines is on the order of at least a decade, perhaps much longer in the United States, and forecasting future supply that far into the future is hazardous.

The lack of mining industry exploration of REE deposits in the last few decades is paralleled by a low level of geological research. The U.S. Geological Survey has demonstrated in related studies that the first step in improving our understanding of REE resources and prospects for further discoveries is to conduct national and global mineral resource assessments. Rare earth elements are one of the commodities under consideration for the next National Resource Assessment, scheduled to begin in 2012. Preliminary work is underway as part of the Minerals at Risk and for Emerging Technologies Project, which will be completed at the end of Fiscal Year 2011.

Table 13. Time Required to Develop Selected Mines Outside of the United States

[These mines selected as close analogs of the mining and metallurgical methods required by a rare earth elements mine. NYA, commercial production not yet achieved]

Mine	Country	Commodity	Discovery date	Permitting completed	Production began	Commercial operations began
Ambatovy	Madagascar	Ni Co	1960	3/2007	Late 2010	NYA
Araxa	Brazil	Nb	1955	1960	1st quarter 1961	1966
Barco Alto	Brazil	Ni	1981	12/2006	3/2010	2011
Bulong	Australia	Ni	1971	9/1996	3/1999	9/1999
Catalão I	Brazil	Nb	1970	1974	1976	1977
Cawse	Australia	Ni Co	1994	9/1996	1/1999	6/2000
Goro	New Caledonia	Ni Co	1982	10/2004	Late 2010	NYA
Murrin Murrin	Australia	Ni Co	1984	5/1996	5/1999	12/2003
Niobec	Canada	Nb	6/1967	11/1973	1/1976	3/1976
Onça Puma	Brazil	Ni	1970s	8/2005	3rd quarter 2010	Early 2011
Raventhorpe	Australia	Ni Co	Late 1960s	3/2004	10/2007	NYA
Urumu Utsumi	Brazil	U	1971	1975	7/1981	1982
Vermelho	Brazil	Ni	1966	7/2005	4th quarter 2012	NYA

THE PRINCIPAL RARE EARTH ELEMENTS DEPOSITS OF THE UNITED STATES

The largest rare earth elements (REE) deposits in the United States are found in carbonatites and alkaline igneous rocks and are concentrated in veins genetically and spatially associated with alkaline igneous intrusions. The association of REE with alkaline igneous rocks also places REE in close association with minerals that host other valuable elements, such as titanium, niobium, phosphorus, and thorium (Van Gosen and others, 2009).

The major REE deposits in the United States are found in

- Carbonatites and alkaline igneous complexes;
- Veins related to alkaline intrusions;
- Some iron ore deposits associated with magmatichydrothermal processes; and
- Stream and beach deposits (placers) derived from the erosion of alkaline igneous terranes.

The principal REE districts in the United States are briefly summarized in this report. More-detailed descriptions of these districts and their mineral deposits are available in the References Cited section. The Mountain Pass REE mine, California, may resume operation within the next 2 years, and some of the districts mentioned in this report have experienced recent exploration activity to evaluate their REE resource potential.

Glossary of Terms

Alkaline igneous rock: A series of igneous rocks that formed from magmas and fluids so enriched in alkali elements that sodium- and potassium-bearing minerals form constituents of the rock in much greater proportion than normal igneous rocks. For detailed discussions of alkaline igneous rocks and their scattered geographic distribution refer to Sorensen (1974) and Woolley (1987).

Carbonatite: A rare, carbonate igneous rock formed by magmatic or metasomatic processes. Most carbonatites consist of 50 percent or more primary carbonate minerals, such as calcite, dolomite, and ankerite. They are genetically associated with, and therefore typically occur near, alkaline igneous rocks. Thorough treatises on carbonatites are provided by Tuttle and Gittins (1966), Heinrich (1980), and Bell (1989).

Epithermal: Mineral veins and ore deposits formed within the Earth's crust from warm water at shallow depths and relatively low temperatures (50–200°C), generally at some distance from the magmatic source.

Hypabyssal: An igneous intrusion that solidified at shallow depths before reaching the Earth's surface.

ALASKA—BOKAN MOUNTAIN

Location: Southern area of Prince of Wales Island, the southernmost island in Alaska. Latitude: 54.91299 N., Longitude: 132.13509 W.; datum: WGS84

Deposit type and basic geology: Several northwest-trending "vein-dike" systems cut linearly through a zoned, peralkaline granite pluton. The vein-dike deposits contain rare earth elements, thorium, and uranium concentrations of several percent each. Individual vein-dike systems extend as much as 2.6 km (1.6 mi) along strike, composed of multiple, subparallel, thin veins that individually rarely exceed 1.5 m (5 ft) in width.

Status: Active, ongoing exploration and assessment of the rare earth elements vein systems in the district by Ucore Rare Metals; its exploration work in the district began in 2007 and has continued into the 2010 field season (http:// www.ucoreraremetals.com/bokan.asp).

Production: Between 1957 and 1971, the Ross-Adams mine was operated by three different companies to fulfill a contract with the Atomic Energy Agency; it produced roughly 85,000 tons (77,000 metric tons) of ore with a grade of about 1 percent uranium oxide and 3 percent thorium oxide.

Estimated resources: An assessment by the U.S. Bureau of Mines (Warner and Barker, 1989) suggested that collectively the vein-dike systems in the district represent a resource of 6.8 million tons (6.2 metric tons) of ore that average 0.264 percent rare earth elements, about one- third of which is yttrium. Recent assay results released by Ucore Rare Metals suggest that in some of the vein-dike deposits the rare earth elements content (dominated by yttrium) can locally exceed 11 percent.

Detailed Discussion

Bokan Mountain is situated near the southern tip of Prince of Wales Island, which is the southernmost island in the Alaska panhandle and which covers an area of about 7–10 km^2 (3–4 mi^2) (figure 4) (Warner and Barker, 1989; Philpotts and others, 1998). The Upper Triassic to Middle Jurassic host rock (Lanphere and others, 1964; De Saint-Andre and others, 1983) is a riebeckite-acmite–bearing peralkaline granite with a crudely circular shape that intruded Paleozoic igneous and sedimentary rocks (Staatz, 1978). The core riebeckite granite porphyry contains subordinate aplitic aegirine granite and is surrounded by an outer annulus composed of predominantly aegirine granite porphyry (Thompson, 1988; Philpotts and others, 1998). Pegmatite-aplites with thorium, rare earth elements (REE), and low levels of gold are also present on Bokan Mountain and were emplaced in contact zones around the intrusive granite (Staatz, 1978; Warner and Barker, 1989; Philpotts and others, 1998). In addition, various dikes cut across all of the rocks near Bokan Mountain, with compositions that include andesite, dacite, basalt, lamprophyre, quartz, monzonite, rhyolite, aplite, and quartz latite (Warner and Barker, 1989). Some of the more felsic dikes contain high levels of accessory Nb, REE, and Th.

The aplitic pegmatites are found throughout the peralkaline granite and range in shape from lensoidal bodies to elongated pods. Examples are radioactive pegmatites exposed in the IML prospects on the east flank of Bokan Mountain, about 1 km (0.6 mi) north-northwest of the Ross-Adams mine (MacKevett, 1963; Warner and Barker, 1989). Most of the pegmatites contain complex mineralogies that include quartz, albite, aegirine, and zircon, with variable amounts of allanite, ilmenite, riebeckite, arsenopyrite, and fluorite (Warner and Barker, 1989). Because of alteration of riebeckite, along the border zone pegmatites typically contain disseminated iron and titanium spinels, as well as magnetite. The cores of the pegmatites consist of milky white massive quartz. The trace element compositions of the pegmatites is equally complex and may contain elevated percentages of Au, Be, Nb, REE, hafnium (Hf), Li, Ta, Sn, Th, U, Y, and Zr. The wall rock also contains a halo that is enriched in these elements, with minerals that include aegirine, sericite, and hematite alteration (Warner and Barker, 1989). Feldspar is largely altered to clay minerals.

In 1955, uranium was discovered in the shear zones and fractures at Bokan Mountain. However, the Ross-Adams mine was the only commercially productive open pit mine in the area. The Ross-Adams mine extracted ore from the Ross- Adams pipe on the Cub claim, which lies along the contact between aegirine syenite and aegirine granite porphyry. An irregularly shaped north-northwest-trending body, the pipe measures 24 m (79 ft) across and was mined along strike for more than 300 m (984 ft) (Thompson, 1988). Between 1957 and 1971, the Ross-Adams mine was operated by three different companies and produced roughly 85,000 tons of ore with a grade of about 1 percent U_3O_8 and 3 percent ThO_2 (Stephens, 1971; Thompson, 1988; Warner and Barker, 1989). Climax Molybdenum operated the mine from 1957 to 1959 and produced 315,000 lb of U_3O_8, and Standard Metals produced another 300,000 lb of U_3O_8 between 1959 and 1964 (table 14). From 1970–1971, Newmont Exploration produced 687,000 lb of U_3O_8, at which time operations were suspended, leaving approximately 365,000 tons of uranium ore unmined (Warner and Barker, 1989).

A private report for the U.S. Department of Agriculture of work conducted by Kent and Sullivan (2004) indicates that a total of 90,700 tons of uranium ore with an unknown grade has been produced at the Ross-Adams mine; in contrast, Cathrall (1994) reports that a total of 209,400 tons of 1 percent uranium oxide (U_3O_8) ore was produced. Several companies, including Standard Metals, Cotter Corp., Union Carbide, Santa Fe Minerals, Dome Minerals, and Humble Oil, conducted exploration drilling on the Bokan Mountain property during 1971–1981 (Kent and Sullivan, 2004). The property is currently (2010) under study by Ucore Rare Metals; the company began an intensive new exploration program in 2007 that continued to the present throughout an area of about 20 mi^2 that includes the Ross-Adams mine property (http://www. ucoreraremetals.com/bokan.asp). Preliminary reports based on these new data suggest an inverse relationship between U and REE; U is located near the margin of the granite complex and more abundant REE more distant from the intrusive complex. However, ore having a high percentage of U is generally also enriched in REE—especially the heavy rare earth elements (HREE), Y, Zr, Be, and Nb. For example, Ucore Rare Metals reports finding about 6 m of core having a grade of 0.26 percent light rare earth elements (LREE) and 3.6 percent HREE in the orebody's I & L vein system.

Table 14. Uranium Production from the Ross-Adams Mine, Alaska

Year	Mining company	Amount mined (tons)	U_3O_8 produced (lb)	Grade (percent U_3O_8)
1957	Climax Molybdenum Corp.	15,000	315,000	1.05
1959–1964	Standard Metals Corp.	15,000	300,000	1.0
1970–1971	Newmont Exploration, Ltd.	55,600	687,000	0.62
	Total	85,600	1,302,000	
	Weighted average			0.76

Source: Warner and Barker, 1989. Stephens (1971) reported 83,000 tons of uranium ore at about 1% U_3O_8. Kent and Sullivan (2004) reported 90,700 tons of ore of unknown grade. Cathrall (1994) reported 209,400 tons of 1 percent U3O8, making the total U_3O_8 produced 4.2 million pounds. lb, pound

Rare earth elements–bearing uraniferous orebodies also lie in the northwest-striking shear zones within the stock granite, where they measure as thick as 3 m (3.3 ft) and as long as 30 m (98 ft) along strike (Thompson, 1988). The main uranium minerals include uranothorite and coffinite in a gangue of quartz and feldspar (Warner and Barker, 1989; Heylmun, 1999). Typically, uranothorite is the dominant ore mineral; it forms yellowish to brownish ovoids that are 0.2 to 2 mm in diameter (Thompson, 1988). Hematite may be found as rims on uranothorite grains or in fine veinlets that extend along microfractures between ovoids. Less than 2 percent of the ore also consists of sulfide species such as pyrrhotite, pyrite, chalcopyrite, galena, sphalerite, marcasite, acanthite, and bornite (Thompson, 1988). Within and adjacent to the orebodies, pervasive hydrothermal albite and minor amounts of chlorite, fluorite, calcite, quartz, sericite, and tourmaline precipitated during wall-rock alteration (Thompson, 1988).

In addition to the mineralized shear zones, U- and REE-mineralized veins and dikes extend out from Bokan Mountain all the way to the West Arm of Kendrick Bay. Most of the veins and dikes are parallel to subparallel (table 15) and crop out on the southeast side of Bokan Mountain, although an exception is the Geiger dike, which is located on the northwest side of the mountain (figure 5). Overall, the veins and dikes tend to bifurcate and anastomose, especially away from the granite stock, which makes them appear more like members of a system rather than individual bodies (Philpotts and others, 1998). On the whole, ore mineralization is confined to microfractures in the dikes or the interstices between larger silicate grains (Warner and Barker, 1989). For example, the I & L vein system strikes west-northwest (table 15) and is located within the peralkaline granite stock on the east side of Bokan

Mountain (MacKevett, 1963; Warner and Barker, 1989). The system measures 2.6 km (1.6 mi) long and is composed of many thin, subparallel veins that individually rarely exceed 1.5 m (5 ft) in width (Staatz, 1978; Warner and Barker, 1989). Of the two main splays in the I & L vein system, the western splay (closest to the Bokan Mountain granite) is hosted by limonite- stained peralkaline granite and pegmatite (Warner and Barker, 1989). The Dotson dike system terminates in the northwest at a right-lateral offset to the I & L vein system and extends about 2 km (1.2 mi) to the southeast; this system of veins has a width around 1 km (0.6 mi) (Staatz, 1978; Philpotts and others, 1998). The Dotson dike system likely extends southeastward beneath the West Arm of Kendrick Bay (Warner and Barker, 1989). The Geiger dike extends northerly from near an aplite outlier of the peralkaline granite to the north shore of South Arm Moira Sound; it is a continuous dike system of one to five or more parallel, radioactive dikes (Warner and Barker, 1989). The Cheri dike system consists of steeply dipping, subparallel, radioactive dikes that cut albitized quartz diorite country rock (Warner and Barker, 1989). In some places, pyroxene-rich masses appear within the dikes whereas magnetite, pyrite, and epidote appear along the edges. Rare fluorite and secondary uranium have also been identified within all of these "vein-dike" systems (Warner and Barker, 1989).

The Upper Cheri dikes are subparallel to and southwest of the Cheri dikes. On the whole, the dikes are very similar to those of the Cheri dike system, with comparable mineralogy, structure, setting, radioactivity, and intrusive characteristics (Warner and Barker, 1989). The Geoduck dikes can be traced southeasterly along a strike of N. 40°–50° W. for about 2.9 km (1.8 mi) (Warner and Barker, 1989) (table 15). Texturally, the dikes are fine to medium grained and granular and banded or cut by veinlets of quartz or opaque minerals. In addition, the Geoduck dike system has, for most of its strike length, chlorite- and epidote-altered pyritic quartz diorite wall rocks (Warner and Barker, 1989).

The veins and dikes are of importance because they contain anomalously high amounts of Be, Nb, REE, Ta, and Hfrich zirconium (table 16). From 1984 to 1987, the U.S. Bureau of Mines investigated several prospects on Bokan Mountain and in the surrounding area, including the mineralized dikes. Collectively, the dikes indicate a resource of 6.8 million tons of ore that average 0.264 percent REE oxides, about one-third yttrium (Warner and Barker, 1989). In addition, the dikes are extensively enriched in yttrium and heavy rare earth elements (HREE) relative to the light rare earths (LREE); yttrium is present at 1,000 times its normal crustal abundance. This composition contrasts with REE deposits elsewhere in the United States and is important because most of the HREE and yttrium in the United States is imported (Warner and Barker, 1989). The most abundant REE are, in order of generally decreasing abundance, Y, Ce, Nd, La, and Sm. Gadolinium, Dy, Ho, Er, and Tm are also present in variable and sometimes noteworthy concentrations (Warner and Barker, 1989). On average, the dikes also contain 0.727 percent zirconium oxide and 0.155 percent niobium oxide, while the amount of thorium and uranium in the dikes is negligible. Trace to minor amounts of other valuable elements are also present, including Be, Ga, Ge, Au, Hf, Pb, Li, Pd, Rb, Ag, Sr, Ta, Sn, V, and Zn.

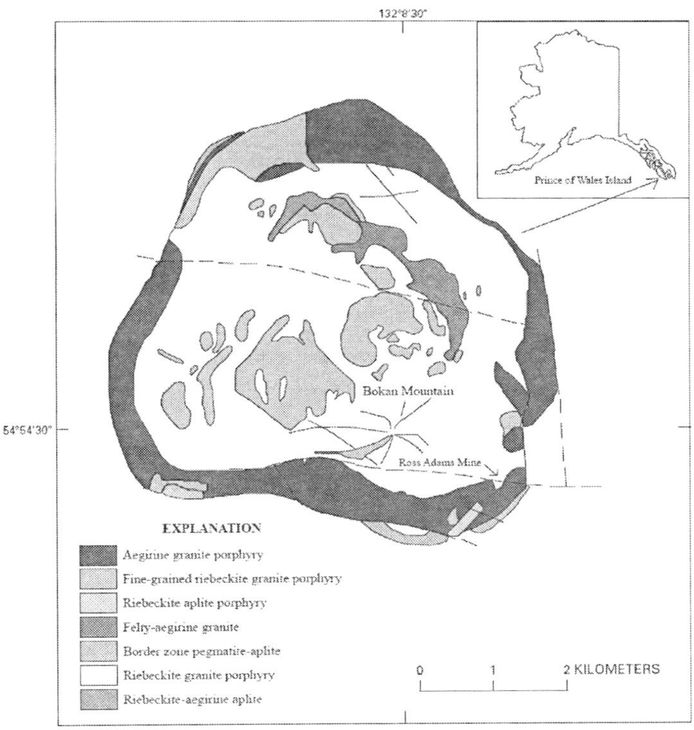

Figure 4. Simplified geologic map of Bokan Mountain, Alaska. Modified from Thompson (1988).

The minerals of the euxenite-polycrase series host most of the Nb found in the dikes, though minor amounts are also contained in columbite (now called ferrocolumbite), aeschynite, and fergusonite (Warner and Barker, 1989). Thalenite, or its alteration product tengerite, contains the observed Y as well as inclusions of xenotime. Other REE are contained within the minerals bastnasite, parisite, synchysite, xenotime, and monazite. Thorium and uranium are present in thorite and uranothorite. Other minerals identified in the dikes include aegirine, barite, biotite, calcite, epidote, fluorite, galena, iron oxides, magnetite, microcline, microperthite, native silver, pyrite, riebeckite, sphalerite, and zircon. Philpotts and others (1998) examined a 3-km (1.9 mi) transect from the margin of the Bokan Mountain peralkaline granite stock along a micro-pegmatite and aplite vein-dike system enriched in Y-REE-Zr-Nb, and they identified minerals such as arfvedsonite, taeniolite, and gittinsite, as well as several other REE-, Zr-, and Nb-bearing phases. By use of various analytical methods, the examined transect was found to be generally enriched in Y and HREE and to have a pronounced negative Eu anomaly, which largely agrees with the results obtained by Warner and Barker (1989) for several dike systems in the Bokan Mountain area.

Table 15. Dimensions of Main Orebodies in the Bokan Mountain District, Alaska

[m, meter; --, not available]

Prospect	Deposit type	Trend	Length (m)	Average width (m)	Estimated depth (m)	Source
Dotson shear zone	Shear zone, fracture controlled	--	--	3.0	--	Warner and Barker (1989).
Ross-Adams pipe	Shear zone, fracture controlled	N-NW	300	24	--	Thompson (1988).
Cheri	Dike	N. 45° W.	1,097	0.9	264	Warner and Barker (1989).
Upper Cheri	Dike	--	366	1	--	Warner and Barker (1989).
Dotson	Dike	--	2,134	0.9	762	Warner and Barker (1989).
Geiger	Dike	N. 15° E. to N. 30° W.	1,707	1.5	762	Warner and Barker (1989).
Geoduck	Dike	N. 40° W. to N. 50° W.	2,896	0.5	762	Warner and Barker (1989.
I & L vein system	Dike	NW	2,600	$\leq$1.5	--	Staatz (1978).

Similar to the mineralogy of the dike systems, the mineralogy of the I & L vein system is complex. In particular, U, Th, and REE are present in several different minerals from different parts of the veins, and they are usually in a gangue dominated by quartz and albite (Staatz, 1978; Keyser and McKenney, 2007). Generally, U is located in Th-bearing uraninite, although it can be found in brannerite in some of the transverse veins (Staatz, 1978). Secondary U minerals, such as kasolite and sklodowskite, have been identified, but they are rare (Keyser and McKenney, 2007). In the northwest part of the vein system, thorite is the main thorium mineral, while allanite is found in the southeastern part of the system as well as in the transverse veins. Other rare earth minerals besides allanite include bastnasite, xenotime, and monazite. However, the distribution of the REE oxides is unusual in that one part of a vein may contain mostly LREE (for example, bastnasite) whereas another part of the same vein has predominantly HREE (for example, xenotime) (Staatz, 1978).

In addition, the I & L vein system contains anomalous amounts of Be, Nb, Zr, Ba, Sr, Sn, Pb, Zn, Cu, and Mo (table 16). Most of the Nb-rich minerals are located in euxenitepolycrase or columbite-tantalite series. (Columbite and tantalite are obsolete names for a mineral series; columbite is now named ferrocolumbite ($Fe^{2+}Nb_2O_6$), which forms two minerals series, one with ferrotantalite ($Fe^{2+}Ta_2O_6$) and the other with manganocolumbite [$(Mn^{2+},Fe^{2+})(Nb,Ta)_2O_6$)].

Commonly, these minerals are associated with zircon as microveinlets. Alternatively, these minerals may replace albite and quartz in the vein matrix. Other minerals include aegirine, barite, biotite, calcite, epidote, fluorite, galena, iron oxides, potassium feldspar, magnetite, pyrite, riebeckite, native silver, and sphalerite (Staatz, 1978).

Many of the dikes in the Bokan Mountain area crop out in heavily forested areas and thus the bedrock is covered by thick vegetation (Warner and Barker, 1989). For instance, the Upper Cheri dike system can be traced only intermittently because of muskeg cover, glacial till, or obscuration by hillside talus where it is open-ended. On the northwest end of the Upper Cheri, the dike system passes under a low-lying creek valley (Warner and Barker, 1989). Heavy soil cover also prohibits detailed mapping in the I & L vein system. In addition, accessibility to the area is limited given that the topography of the Bokan Mountain area ranges from moderately steep to precipitous (Keyser and McKenney, 2007). Although deep-water marine access is available to and from Ketchikan and Prince Rupert by way of Kendrick Bay and Moira Sound, accessing more remote sections of the area must be achieved by helicopter, boat, or on foot. In spite of these limitations, the mineralization could have considerable economic potential.

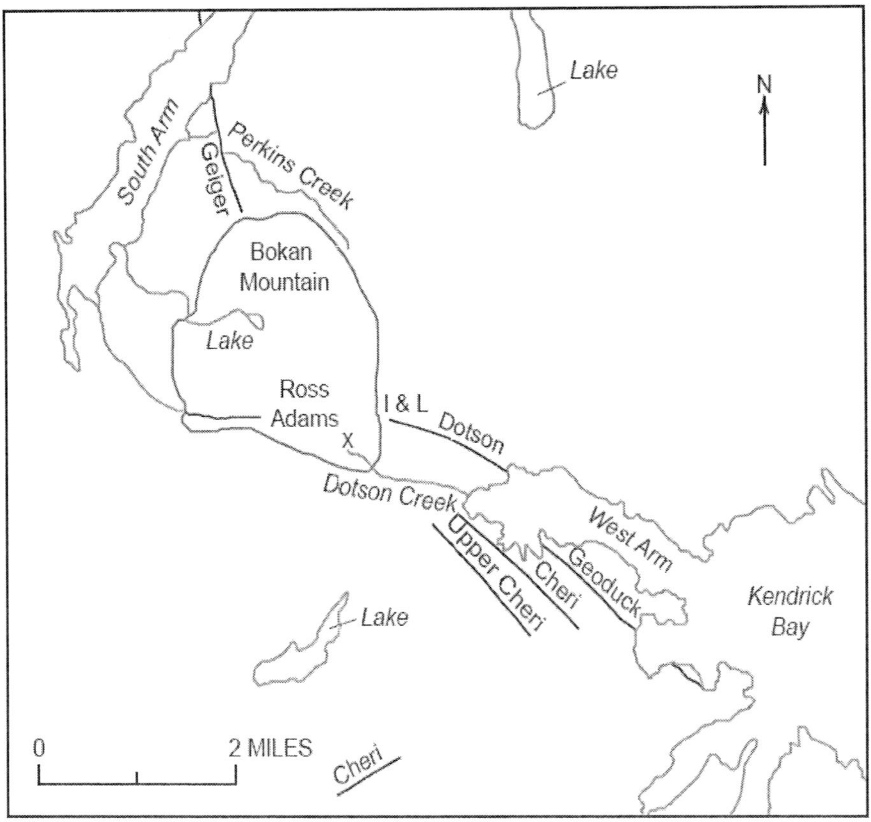

Figure 5. Map of major vein and dike systems associated with Bokan Mountain, Alaska. Modified from Heylmun (1999).

ALASKA—SALMON BAY

Location: Northeast shore of Prince of Wales Island, the southernmost island in Alaska. Latitude: 56.31915 N., Longitude: 133.17145 W.; datum: WGS84

Table 16. Resource Estimates for Main Prospects on Bokan Mountain, Alaska, and Surrounding Property
[%, percent; --, not available. Source of data: Warner and Barker (1989) and Keyser and McKenney (2007)]

Prospect	Resource (tons)		U_3O_8 (percent)	Nb_2O_5 (percent)	ThO_2 (percent)	Y_2O_3 (percent)	ZrO_2 (percent)	REO (percent)	BeO (percent)	Ta_2O_5 (percent)
	Indicated	Inferred								
Cheri[1]	--	73,000	0.012	0.089	0.025	0.095	0.320	0.281	0.025	0.01
Cheri[1]	--	458,000	0.020	0.175	0.042	0.222	0.650	0.461	0.03	0.01
Cheri[1]	4,443,000	--	0.014	0.123	0.032	0.180	0.410	0.352	0.025	0.01
Upper Cheri	481,000	--	0.014	0.099	0.025	0.159	0.460	0.411	--	--
Dotson shear zone	--	2,039,000	0.009	0.083	0.071	0.113	0.009	0.132	--	--
Dotson dike1	8,490,000	--	0.011	0.103	0.052	0.138	0.260	0.21	0.16	0.026
Geiger[1]	--	2,450,000	0.021	0.219	0.009	0.168	2.430	0.458	--	0.014
Geiger[1]	4,693,000	--	0.021	0.219	0.168	2.430	0.458	--	--	0.014
Geiger[1]	2,600,000	--	0.008	0.062	0.009	0.163	1.000	0.149	--	--
Geoduck[1]	--	1,378,000	0.015	0.148	0.031	0.374	0.640	0.375	0.028	--
Geoduck[1]	9,528,000	--	0.012	0.112	0.021	0.200	0.390	0.319	0.028	--
I & L	--	21,000	0.094	0.186	0.114	--	0.094	--	--	--
I & L	--	23,000	--	0.096	--	--	--	--	--	--
I & L	--	50,000	--	0.100	--	--	--	--	--	--
I & L	--	6,000	--	0.300	--	--	--	--	--	--
I & L	37,000	--	--	0.100	--	--	--	--	--	--
I & L	5,000	--	--	0.300	--	--	--	--	--	--
ILM	586,000	--	0.017	0.176	--	0.083	3.340	0.353	--	--
Ross-Adams mine	--	365,000	0.17	--	0.460	0.400	0.320	0.32	--	--
Sunday Lake	27,000	--	0.59	0.070	3.640	1.030	0.380	0.329	--	--
Total	30,890,000	6,863,000								

[1]Trace percentages of Ge, Hf, Pb, Zn, and Au also present.

Deposit type and basic geology: Short, irregular, and lenticular veins of radioactive carbonate-hematite crop out along the coast for about 13 km (8 mi). Some veins can be traced for more than 91 m (300 ft) between the low-tide line and forest cover. The veins cut the Salmon Bay greywacke, a thick formation of Silurian age. On average, the veins are 5–8 cm (2–3 in.) wide, though they normal range from less than 2 cm (1 in.) to as much as 0.76 m (2.5 ft). A few veins reach 1.5–3 m (5 to 10 ft) in width.

Status: Apparently little geologic work has been done in this area since the 1950s.

Production: No past production.

Estimated resources: The average of seven samples taken from one of the more radioactive veins was 0.034 percent equivalent uranium (eU) or 0.156 percent equivalent thorium (eTh), which equates to 0.178 percent equivalent ThO_2 ($eThO_2$) (Houston and others, 1955). The fluorine- rich carbonates from the highest-grade rare earth elements vein yield an average content of 0.79 percent rare earth oxides. Because of the short, lenticular, and irregular nature of the veins, the average grade or total reserves was not calculated.

Detailed Discussion

Reconnaissance for radioactive deposits in southeastern Alaska in 1952 identified radioactive minerals in the vicinity of Salmon Bay, Alaska, located on the northeastern shoreline of Prince of Wales Island. Short, irregular, and lenticular veins of radioactive carbonate-hematite crop out along the coast for about 13 km (8 mi), roughly 5 km (3 mi) northwest and 8 km (5 mi) southeast of Salmon Bay (Houston and others, 1955). Some of the veins can be traced, however, for more than 91 m (300 ft) between the low-tide line and forest cover. The veins cut the Salmon Bay greywacke, a thick formation of Silurian age that ranges in color from reddish brown to grayish green (Houston and others, 1955). On average, the veins are 5–8 cm (2–3 in.) wide, though they normally range from less than 2 cm (1 in.) to as many as 0.76 m (2.5 ft). A few veins reach 1.5–3 m (5 to 10 ft) in width.

The predominant minerals in the veins are dolomite-ankerite and alkali feldspar, with lesser amounts of hematite, pyrite, siderite, magnetite, quartz, chalcedony, and chlorite (Houston and others, 1955). Other minerals identified include parisite, bastnasite, muscovite, fluorite, apatite, thorite, zircon, monazite, epidote, topaz, garnet, chalcopyrite, and marcasite. The radioactivity in the veins is caused by thorite and monazite, both of which contain thorium. The fluorcarbonates parisite and bastnasite are found in nonradioactive carbonate- hematite veins, which are also located along the coast and are wider than the radioactive veins. Of the two fluorcarbonates, parisite is more abundant and appears to be a late-stage mineral that fills in small vugs or was deposited along fractures in the host carbonate vein (Houston and others, 1955).

Seven samples taken from one of the more radioactive veins averaged 0.034 percent equivalent uranium (eU) or 0.156 percent equivalent thorium (eTh), which equates to 0.178 percent equivalent ThO_2 ($eThO_2$) (Houston and others, 1955). The fluorcarbonates from the highest-grade rare earth vein yield an average content of 0.79 percent rare earth oxides. Because of the short, lenticular, and irregular nature of the veins, no calculation of the average grade or total reserves was attempted. However, analyses of samples from the Paystreak vein on Pitcher Island were sufficient to enable preliminary estimates. Houston and others (1955) reported approximately 68.6 lb (31 kg) of Th or 78.1 lb (35.4 kg) of ThO_2 per foot (0.3 m) of depth for the 100-ft (30.5-m) portion of the vein sample.

Additional exploration of the Salmon Bay deposit is necessary to more fully evaluate the economic potential of this resource, as little work has been done in this area since the 1950s.

CALIFORNIA—MOUNTAIN PASS DEPOSIT AND MINE

Location: Northeastern corner of San Bernardino County, California. Latitude: 35.478 12 N., Longitude: 115.53068 W.; datum: WGS84

Deposit type and basic geology: A massive carbonatite called the Sulphide Queen body forms the core of the Mountain Pass igneous complex and hosts the bulk of the rare earth elements resources in the district. This carbonatite body has an overall length of 730 m (2,395 ft) and average width of 120 m (394 ft). The typical ore contains about 10–15 percent bastnasite (the ore mineral), 65 percent calcite or dolomite (or both), and 20–25 percent barite, plus other minor accessory minerals (Castor and Nason, 2004). The

Sulphide Queen carbonatite body is the largest known mass of high-grade rare earth elements ore in the United States.

Status: Molycorp ceased its mining of the Mountain Pass rare earth elements deposit in 2002 when its permit expired. In 2009, Molycorp announced its intentions to resume mining at Mountain Pass by the year 2012.

Production: Rare earth elements were mined in the district beginning in 1952, with nearly continuous production until 2002. The mine's peak output, around 1990, was 20,000 metric tons/year (22,000 tons/year) of rare earth elements oxides.

Estimated resources: Molycorp estimates that the remaining deposit holds 20 to 47 million metric tons (22 to 52 million tons) of ore with an estimated average grade of 8.9 percent rare earth elements oxide.

Detailed Discussion

The Mountain Pass deposit sits near the eastern edge of the Mohave Desert in the northeastern corner of San Bernardino County, California. It lies just north of Interstate Highway 15 near Mountain Pass, about 60 mi (96 km) southwest of Las Vegas, Nevada (figures. 6 and 7).

The Mountain Pass deposit is commonly recognized as the largest known rare earth elements (REE) resource in the United States, with current reserves estimated to be greater than 20 million metric tons of ore with an average grade of 8.9 percent rare earth elements oxide (Castor and Hedrick, 2006). A massive carbonatite called the Sulphide Queen body forms the core of the Mountain Pass igneous complex and hosts the bulk of the REE mineral resources in the district. This carbonatite body has an overall length of 730 m (2,395 ft) and average width of 120 m (394 ft) (Olson and others, 1954). The typical ore contains about 10–15 percent bastnasite (the ore mineral), 65 percent calcite or dolomite (or both), and 20–25 percent barite, plus other minor accessory minerals (Castor and Nason, 2004). The Sulphide Queen carbonatite body is the largest known mass of high-grade REE ore in the United States. Light REE are preferentially concentrated in the Mountain Pass ore (Castor, 2008).

Molycorp ceased its mining of the Mountain Pass REE deposit in 2002 when its permit expired. However, in 2009, Molycorp announced its intentions to resume mining at Mountain Pass by the year 2012. The mine's open pit—

inactive since 2002—covers about 22 hectares (55 acres) of area and is about 122 m (400 ft) deep. In July 2009, Molycorp reached agreement with Arnold Magnetic Technologies Corp. of Rochester, New York, to make permanent magnets using REE mined at Mountain Pass (Mining Engineering, 2009). Molycorp announced that "Plans call for mining to resume at Mountain Pass by 2012, at the rate of about 972 t/d [972 metric tons per day; 1,000 tons per day] of ore, enough to produce 20 kt [20,000 metric tons; 22,000 tons] of rare earth oxides for sale each year" (Mining Engineering, 2009, p. 8); it has received approval to double its output volume with time. The mine's peak output 20 years ago was 20,000 metric tons per year of rare earth oxides (Mining Engineering, 2009). Molycorp estimates that the remaining deposit holds 20 to 47 million metric tons (22 to 52 million tons) of ore (Mining Engineering, 2009).

The Sulphide Queen carbonatite stock and nearby carbonatite dikes are associated with Proterozoic, potassium-rich igneous rocks—biotite shonkinite, hornblende and biotite syenite, and granite—that intruded Precambrian metamorphic and igneous rocks (Olson and others, 1954; Castor, 2008). The Sulphide Queen body was originally mapped as three rock types with local variations—gray calcite-barite rock (figure 8), ferruginous dolomitic rock, and silicified carbonate rock (Olson and others, 1954). All phases of the stock contain bastnasite. Age determinations indicate that the Sulphide Queen carbonatite was emplaced 1375±5 million years ago (DeWitt and others, 1987), about 25–35 million years after the alkaline igneous intrusions in the district (Castor, 2008).

In addition to the massive Sulphide Queen carbonatite stock, several carbonatite dikes are exposed in the Mountain Pass district. These dikes include the Birthday veins north and northwest of the Sulphide Queen mass and other carbonatite dikes southeast of the stock. The Sulphide Queen stock and the carbonatite dikes are exposed across an area of less than 1 mi^2 in the district. The carbonatite dikes range from a few inches to about 6 m (20 ft) in thickness and can be exposed for as much as 122 m (400 ft) along strike (Olson and others, 1954). Like the Sulphide Queen stock, the carbonatite dikes are abundant in calcite (as much as 90 percent of the dike) and barite (as much as 30 percent); but in contrast to the stock, individual dikes may or may not contain bastnasite. Accessory minerals include siderite, quartz, fluorite, galena, pyrite, apatite, crocidolite, wulfenite, biotite, thorite, hematite, magnetite, goethite, and potassium feldspar (Olson and others, 1954; Castor and Nason, 2004; Castor, 2008). The carbonatite dikes range in total REE oxide content from 2.03 to 18.64 percent (Olson and others, 1954; Staatz and others, 1980).

The REE content of the Mountain Pass district was first discovered in April 1949, when a couple of uranium prospectors noted modest radioactivity on their Geiger counter on Sulphide Queen hill and at the Birthday vein, located 1,219 m (4,000 ft) to the northwest (radioactivity reflected the thorium content of the carbonatite). They grabbed samples of radioactive rock and took it to the U.S. Bureau of Mines office in Boulder City, Nevada. Analyses of the samples by the Bureau of Mines, confirmed by laboratories of the USGS, found that the rock was rich in bastnasite—a rare earth–carbonatefluorine mineral—which subsequently became the primary ore mineral of Mountain Pass. Later in 1949, the prospectors filed claims on the Birthday vein system. In November of 1949, the USGS initiated a high-priority field study of the Mountain Pass district, which mapped, described, and sampled the district in detail; this field work led to the discovery of the massive Sulphide Queen carbonatite stock (Shawe, 1953; Olson and others, 1954).

Figure 6. Google Earth image of the Mountain Pass mining district, California. Molycorp's open pit mine—inactive since 2002—is at the center of this view; the pit covers about 55 acres (22 hectares) and is about 400 ft (122 m) deep. (Image used with permission of Google).

Mappers collected 59 outcrop samples of the Sulphide Queen stock, which in total showed an average content of 6.9 percent REE oxides. Molybdenum Corporation of America (later "Molycorp") purchased the claims from the prospectors in February 1950 and prepared plans to mine these bastnasite deposits. In 1952, Molybdenum Corporation of America began mining of the surface outcrops and built small processing facilities in the district. The early prospecting and discovery history of the Mountain Pass district is thoroughly described by D.F. Hewett, as the foreword to Olson and others (1954).

Figure 7. Northwest-facing view of Mountain Pass district, California, about 1997, viewed from the Mineral Hill area south of Interstate Highway 15. An outcrop of ultrapotassic rock is in the right foreground. (Photograph by Stephen B. Castor, Nevada Bureau of Mines and Geology; used with permission).

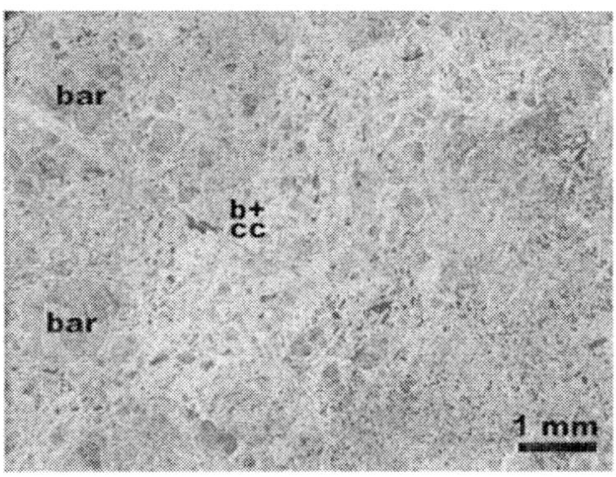

Figure 8. Dolomitic carbonatite ("beforsite," described by Castor, 2008) of the Sulphide Queen orebody, Mountain Pass district, California. This sample is very high grade ore; it contains more than 12 percent rare earth oxide. bar, barite phenocrysts; b+cc, fine-grained bastnasite mixed with calcite; brown to yellow mineral is dolomite. (Photograph by Stephen B. Castor, Nevada Bureau of Mines and Geology; used with permission).

Although the report is more than half a century old and was written prior to large-scale mining in the district, the USGS report of Olson and others

(1954) remains the most comprehensive published report on the geology of the Mountain Pass district. The geologic mapping, lithologic descriptions, and mineralogy described in the report have been proven through the several decades of development to be an accurate geologic accounting of the district. The results of more recent geologic and geochemical research in the district, which includes subsurface information, are detailed by Castor (2008).

California—Music Valley Area

Location: Sixteen km (10 mi) southeast of Twentynine Palms in Riverside County, southern California. Latitude: 33.98423 N., Longitude: 115.93254 W.; datum: WGS84

Deposit type and basic geology: The xenotime deposits of the Music Valley area are situated within the Pinto Gneiss of probable Precambrian age. Biotite-rich zones in the gneiss can contain abundant orange xenotime grains, commonly forming 10–15 percent of the biotite zones and locally comprising about 35 percent xenotime.

Status: A reported deposit with apparently no current exploration activity.

Production: These deposits had no past production. Small- scale exploration of these deposits during the late 1950s investigated their radioactivity.

Estimated resources: No resource estimate has been made for this district. In 1957, rock chip samples were collected in radioactive, biotite-rich intervals of the Pinto Gneiss that had been exposed in exploration prospect pits. Results from five samples found 3.5 to 8.8 weight percent yttrium.

Detailed Discussion

Music Valley lies in the Pinto Mountains about 16 km (10 mi) southeast of Twentynine Palms in Riverside County, southern California, and just to the northeast of Joshua Tree National Park. During 1949 and 1952, reconnaissance ground and airborne radiometric prospecting by the USGS in conjunction with the U.S. Atomic Energy Commission led to the discovery of radioactivity anomalies in the Music Valley area. This discovery led to local exploration efforts to find the source of the radioactivity. By 1959, only traces of uranium had been found in this area, but localized concentrations of xenotime suggest that the radioactivity originated in thorium rather than uranium.

All of the xenotime deposits of the Music Valley area lie within the Pinto Gneiss of probable Precambrian age (Evans, 1964), the oldest rock unit exposed in this area. The Pinto Gneiss consists of roughly equal amounts of quartz and plagioclase feldspar, and it averages approximately 35 percent biotite (Evans, 1964). Accessory minerals present in trace amounts in the gneiss include sericite, apatite, magnetite, zircon, and sphene, and local monazite, actinolite, orthoclase, microcline, perthite, and muscovite. Biotite-rich zones in the gneiss can contain abundant orange xenotime grains that commonly form 10–15 percent of the biotite zones and locally comprise about 35 percent xenotime (Evans, 1964).

In the Music Valley area, xenotime concentrations are distributed throughout a northwest-trending zone that is about 4.8 km (3 mi) in width by 9.7 km (6 mi) in length. According to Evans (1964, p. 10): "Xenotime is almost entirely confined to the Pinto Gneiss where it is irregularly distributed and only locally concentrated in sufficient quantity to give an abnormal radioactive anomaly. It nearly always occurs in biotite-rich lenses, pods, and folia."

In 1957, rock-chip samples were collected in radioactive, biotite-rich intervals of the Pinto Gneiss that had been exposed in exploration prospect pits. Five rock-chip samples collected from the most radioactive prospect, the U-Th deposit, found the following concentrations (Evans, 1964, table 2, p. 21):

Element	Concentration (wt percent)
Y	3.5–8.8
La	0.26–0.34
Ce	0.47–0.94
Nd	0.24–0.41
Dy	0.22–0.41
Yb	0.46–0.75
Th	0.31–0.49

To better evaluate the potential rare earth elements resources within the Music Valley area requires additional detailed mapping, sampling and analyses, and perhaps core drilling.

COLORADO—IRON HILL CARBONATITE COMPLEX

Location: Located near the small town of Powderhorn, about 35 km (22 mi) south-southwest of Gunnison, Colorado. Latitude: 38.253 19 N., Longitude: 107.05328 W.; datum: WGS84

Deposit type and basic geology: A massive carbonatite stock forms the core of the Iron Hill carbonatite complex. The carbonatite stock is enriched in rare earth elements, niobium, and thorium; the adjacent pyroxenite unit is enriched in these elements also and in substantial amounts of titanium.

Status: Since 1990, Teck Resources Ltd. has owned many of the patented claims within this intrusive complex. Its interests have focused on the substantial titanium resource within the pyroxenite unit of the complex. Currently (2010), it appears that Teck Resources is not actively conducting work at this property.

Production: No mineral resources have been produced from this intrusive complex, despite its varied and substantial mineral resources (Van Gosen and Lowers, 2007).

Estimated resources: The U.S. Geological Survey (Staatz and others, 1979) estimated that the carbonatite stock of Iron Hill contains 655.6 million metric tons (722.7 million tons) of carbonatite. On the basis of the averaged analytical results of 28 samples of the carbonatite stock—0.4 percent total rare earth elements oxides and 0.004 percent thorium oxide—Staatz and others (1979) calculated potential reserves within the stock of 2.6 million metric tons (2.865 million tons) of rare earth elements oxides and 28,190 metric tons (31,080 tons) of thorium oxide. Applying an average grade of 0.057 percent niobium oxide, Staatz and others (1979) estimated a reserve of 374,000 metric tons (412,000 tons) of niobium oxide in the carbonatite stock of Iron Hill. For an area of the pyroxenite unit, Teck Resources reported that "mineable proven, probable and possible reserves***are 41.8 million tonnes [46 million tons] grading 13.2% TiO_2 [titanium oxide] within an open-ended global proven, probable and possible geologic resource of 1.6 billion tonnes [1.8 billion tons] grading 10.9% TiO_2" (Shaver and Lunceford, 1998, p. 63).

Detailed Discussion

The Iron Hill carbonatite complex is exposed for 31 km^2 (12 mi^2) near the small town of Powderhorn, about 35 km (22 mi) south-southwest of Gunnison, Colorado. The intrusion is alkaline with a prominent carbonatite stock at its core. This intrusive complex is noteworthy because of its classic geology and its mineral resource potential (Van Gosen and Lowers, 2007). This intrusive complex was described by Olson and Hedlund (1981, p. 5) as "the best example of the carbonatite-alkalic rock association in the United States and is one of the outstanding occurrences in the world, comparable to many of the classic areas in Africa and other continents." The primary rock types of the complex are, from oldest to youngest, pyroxenite, uncompahgrite, ijolite, nepheline syenite, and carbonatite (Olson, 1974; Hedlund and Olson, 1975; Olson and Hedlund, 1981; Armbrustmacher, 1983). Substantial titanium concentrations have been measured in the pyroxenite unit, which is thought to host the largest titanium (Ti) resource in the United States (Thompson, 1987; Shaver and Lunceford, 1998; Van Gosen and Lowers, 2007). The carbonatite stock is enriched in rare earth elements (REE), niobium (Nb), and thorium (Th); the pyroxenite unit is also enriched in these elements plus vanadium (V). Thus, it may be economic to extract several resources from this complex with a well-coordinated mine and mill plan. Thus far, none of these resources has been developed at Iron Hill.

A dolomitic carbonatite stock was the last major igneous phase of the Iron Hill intrusive complex. The stock forms Iron Hill (figure 9) and the ridge to its northwest, and it is exposed throughout an area of about 3.7 km (2.3 mi) long by 1.9–0.8 km (1.2–0.5 mi) wide, making it the largest exposed carbonatite mass in the United States. Staatz and others (1979) estimated that the carbonatite stock of Iron Hill contains 655.6 million metric tons (722.7 million tons) of carbonatite. On the basis of the averaged analytical results of 28 samples of the carbonatite stock—0.4 percent for total rare earth oxides and 0.004 percent ThO_2—Staatz and others (1979) calculated potential reserves within the stock of 2.6 million metric tons (2.865 million tons) of rare earth elements oxides and 28,190 metric tons (31,080 tons) of ThO_2. Recent sampling of the Iron Hill carbonatite stock by Van Gosen (2008) found median values (from 13 samples) of 0.19 percent total rare earth oxides (table 17) and 0.0035 percent ThO_2; this result represents estimated resources within the stock of about 1.22 million metric tons (1.34 million tons) of rare earth elements oxides and about 23,000 metric tons (25,300 tons) of ThO_2. (Median values were used for the Van Gosen (2008) data because a few individual results exceeded the upper analytical limit of detection for an element.)

Applying an average grade of 0.057 percent Nb_2O_5, Staatz and others (1979) estimated a reserve of 374,000 metric tons (412,000 tons) of Nb_2O_5 in the carbonatite stock of Iron Hill. Armbrustmacher and Brownfield (1979) found 0.003 to 0.2 weight percent Nb (niobium) in 28 samples of the carbonatite stock. Van Gosen (2008) found median values of 0.0595 weight percent Nb_2O_5 from 13 samples of the stock, suggesting an estimated resource of 390,000 metric tons (430,000 tons) of Nb_2O_5.

As noted earlier, the pyroxenite unit of the Iron Hill complex is enriched in titanium and likely is the largest titanium resource in the United States. In 1968, Buttes Gas & Oil Company purchased the properties of the Iron Hill intrusive complex. It focused its exploration and development primarily on titanium resources in the perovskite-rich pyroxenite on the northeast side of the Cimarron fault (Thompson, 1983). Exploration by Buttes Gas & Oil continued at the site into the 1 980s. A 1976 newspaper article in the *Denver Post* (February 25, 1976, p. 31) reported that company officials stated that their studies had identified 419 million tons (380 metric tons) of reserves averaging 12 percent TiO_2. Thompson (1987, p. 27) noted, "In 1976, Kaiser Engineers, Inc. prepared a computer ore reserve analysis indicating a reserve of 390,000,000 tons (350 million metric tons) assaying 11.5 percent TiO_2. Since 1976, additional drilling has increased the reserve to at least 500 million tons (450 million metric tons) of about the same grade." Thompson (1983, 1987) summarizes the exploration work at Iron Hill during the 1 970s and 1 980s by Buttes Gas & Oil and its subsidiary companies, and he describes the processing steps that were being considered to most effectively extract titanium from the rock.

In 1990, Teck Resources Ltd. entered into a joint venture partnership with Buttes Gas & Oil Company to explore the titanium resources of the Iron Hill intrusive complex. In 1994, Teck Resources purchased 100 percent interest in the properties. Since then (and by 2010), Teck Resources delineated an orebody within the pyroxenite rock in the northeastern area of the intrusive complex, calculated reserve estimates, and conducted mineral processing, marketing, and environmental baseline studies to evaluate the economic viability of developing these titanium resources (Shaver and Lunceford, 1998). Teck Resources reported that "mineable proven, probable and possible reserves... are 41.8 million metric tons (46 million tons) grading 13.2 percent TiO_2 within an open-ended global proven, probable and possible geologic resource of 1.6 billion metric tons (1.8 billion tons) grading 10.9 percent TiO_2" (Shaver and Lunceford, 1998, p. 63).

More recently, Van Gosen (2008) collected 24 widely scattered near-surface samples of the pyroxenite unit in the northern part of the intrusive complex (figure 10). Results showed a high concentration of 5.74 percent Ti (titanium) with a median value of 3.2 percent Ti content. For comparison, Best (1982, p. 615) reported that the titanium content of typical pyroxenite is approximately 0.88 percent. Also, Upton (1967, p. 283) reported that the Iron Hill pyroxenite contained the highest titanium concentrations among his example chemistries of alkaline pyroxenites worldwide.

Similar to the central carbonatite stock, the pyroxenite unit at Iron Hill also is enriched in rare earth elements (table 17), niobium, and thorium and, additionally, vanadium. Pyroxenite samples collected by Van Gosen (2008) contained median values of 0.143 percent total rare earth elements oxide content (table 2), 0.036 percent Nb_2O_5, 0.0025 percent ThO_2, and 0.046 percent V_2O_5.

Figure 9. Northwest- facing view of Iron Hill, Gunnison County, southwestern Colorado. Iron Hill is formed by a massive carbonatite stock that forms the center of an alkaline intrusive complex. This complex hosts many mineral resources, including titanium, niobium, rare earth elements, and thorium (Van Gosen and Lowers, 2007). The carbonatite stock is estimated to consist of 655.6 metric tons of carbonatite containing 2.6 million metric tons of rare earth elements oxides, 28,200 metric tons of thorium oxide, and 373,700 metric tons of niobium oxide (Staatz and others, 1979, p. 30).

Table 17. Median Concentrations of Rare Earth Elements in Samples of Carbonatite Stock and Pyroxenite Unit, Iron Hill Carbonatite Complex, Colorado

[Rare earth elements listed in order of increasing atomic number; yttrium (Y) is included with these elements because it shares chemical and physical similarities with the lanthanides. ppm, parts per million; wt percent, weight percent. Data from Van Gosen (2008)]

Carbonatite stock (13 samples)				Pyroxenite unit (24 samples)			
Element	Median value (ppm)	Oxide	Oxide equivalent (wt percent)	Element	Median value (ppm)	Oxide	Oxide equivalent (wt percent)
La	344	La_2O_3	0.040	La	264	La_2O_3	0.031
Ce	681	Ce_2O_3	0.080	Ce	508	Ce_2O_3	0.060
Pr	89.7	Pr_2O_3	0.010	Pr	60.2	Pr_2O_3	0.007
Nd	337	Nd_2O_3	0.039	Nd	227	Nd_2O_3	0.026
Sm	47.8	Sm_2O_3	0.006	Sm	39.1	Sm_2O_3	0.005
Eu	11	Eu_2O_3	0.001	Eu	10.6	Eu_2O_3	0.001
Gd	33.7	Gd_2O_3	0.004	Gd	31.1	Gd_2O_3	0.004
Tb	3.72	Tb_2O_3	0.0004	Tb	3.64	Tb_2O_3	0.0004
Dy	8.04	Dy_2O_3	0.0009	Dy	14.1	Dy_2O_3	0.002
Ho	1.10	Ho_2O_3	0.0001	Ho	2.04	Ho_2O_3	0.0002
Er	2.63	Er_2O_3	0.0003	Er	4.48	Er_2O_3	0.0005
Tm	0.29	Tm_2O_3	0.00003	Tm	0.45	Tm_2O_3	0.00005
Yb	1.6	Yb_2O_3	0.0002	Yb	2.5	Yb_2O_3	0.0003
Lu	0.17	Lu_2O_3	0.00002	Lu	0.29	Lu_2O_3	0.00003
Y	27.9	Y_2O_3	0.004	Y	47.8	Y_2O_3	0.006
Total	1,590		0.186		1,215		0.143

The high REE resource estimates at Iron Hill, Colorado, are biased by the very large volume of the host rocks—the carbonatite and the pyroxenite unit—which are exposed throughout very large areas and extend at depth for at least several hundreds of meters. Also, note that REE concentrations at Iron Hill—about 0.14–0.19 percent REE oxide—are substantially lower than the concentrations in the Mountain Pass carbonatite deposit (California), which reportedly averages 8.9 percent rare earth oxide.

Figure 10. Outcrop of pyroxenite unit in the northern part of the Iron Hill carbonatite complex, southwestern Colorado. Analyses of 24 widely scattered samples of this pyroxenite unit collected throughout the complex contained median abundances of 5.3 percent titanium oxide, 0.06 percent cerium oxide, 0.143 percent total rare earth elements oxides, 0.036 percent niobium oxide, 0.0025 percent thorium oxide, and 0.046 percent vanadium oxide (Van Gosen, 2008).

COLORADO—WET MOUNTAINS AREA

Location: In the Wet Mountains and surrounding area in Fremont and Custer Counties of south-central Colorado. Latitude: 38.16695 N., Longitude: 105.21388 W.; datum: WGS84

Deposit type and basic geology: Thorium (Th) and rare earth elements (REE) are noted in veins, syenite dikes, fracture zones, and carbonatite dikes (Armbrustmacher, 1988) associated with three Cambrian alkaline complexes (Olson and others, 1977) that intruded the surrounding Precambrian terrane. Thorium-REE–mineralized veins and fracture zones, which are distal to the three alkaline intrusive complexes, have the highest economic potential for thorium and rare earth elements resources. The thorium-REE veins and fracture zones are linear features, typically 1–2 m (3.3–6.6 ft) thick, but a few are as much as 15 m (49 ft) thick. Some individual thorium veins can be traced in outcrop for 1.5 km (0.9 mi) and some radioactive fracture zones as much as 13 km (8 mi). Most of these vein and fracture-zone deposits lie within a 57 km^2 (22 mi^2) tract of Precambrian gneiss and migmatite located south and southeast of a quartz syenite complex at Democrat Creek. Christman and others (1953, 1959) mapped nearly 400 veins in this area.

Status: No apparent exploration activity is underway in this district at present (2010). Many of the prospective vein and fracture-zone deposits occur on private lands.

Production: No thorium or rare earth elements have been produced from the district. Modest exploration activity, apparently during the 1950s, prospected the radioactive veins in the district; the prospecting included shallow trenching.

Estimated resources: The U.S. Geological Survey (Armbrustmacher, 1988) estimated that the vein and fracture zone deposits of the Wet Mountains area contain the following resources:

- Thorium oxide (ThO_2)
  Reserves of 58,200 metric tons (64,200 tons);
  Probable potential resources of 145,600 metric tons (160,500 tons);
- Total light rare earth elements
  Reserves of 26,600 metric tons (29,300 tons);
  Probable potential resources of 66,500 metric tons (73,270 tons)
- Total heavy rare earth elements
  Reserves of 17,700 metric tons (19,540 tons);
  Probable potential resources of 44,300 metric tons (48,850 tons).

(This estimate was based on average concentrations of 0.46 percent ThO_2, 0.21 percent total light rare earth elements oxides, and 0.14 percent total heavy rare earth elements oxides.)

Detailed Discussion

This thorium-rare earth elements (REE) district, located in Fremont and Custer Counties of south-central Colorado, may be comparable in thorium and REE resources to the Lemhi Pass district of Idaho-Montana. Thorium-REE deposits are exposed throughout an area of about 60 km (37 mi) north to south by 24 km (15 mi) west to east. Thorium and REE are found in veins, syenite dikes, fracture zones, and carbonatite dikes (Armbrustmacher, 1988) associated with three Cambrian alkaline complexes (Olson and others, 1977) that intruded the surrounding Precambrian terrane. These three alkaline complexes are the McClure Mountain Complex (Shawe and Parker, 1967; Armbrustmacher, 1984), the Gem Park Complex (Parker and Sharp, 1970), and the complex at Democrat Creek (Armbrustmacher, 1984). The thorium-REE-mineralized veins and fracture zones, which are distal to the three alkaline intrusive complexes, have the highest economic potential for thorium and REE resources.

On the basis of 201 samples of veins and fracture zones, the USGS (Armbrustmacher, 1988) estimated that the vein and fracture zone deposits of the Wet Mountains area contain the following resources:

- Thorium oxide (ThO_2)
  Reserves of 58,200 metric tons (64,200 tons);
  Probable potential resources of 145,600 metric tons (160,500 tons);
- Total light rare earth elements
  Reserves of 26,600 metric tons (29,300 tons);
  Probable potential resources of 66,500 metric tons (73,270 tons);
- Total heavy rare earth elements
  Reserves of 17,700 metric tons (19,540 tons); and
  Probable potential resources of 44,300 metric tons (48,850 tons).

This estimate incorporates average concentrations of 0.46 percent ThO_2, 0.21 percent total light REE oxides, and 0.14 percent total heavy REE oxides.

The thorium-REE veins and fracture zones are linear features, typically 1–2 m (3.3–6.6 ft) thick, but a few are as much as 15 m (49 ft) thick (figure 11). Some individual thorium veins can be traced in outcrop for 1.5 km (0.9 mi) and some radioactive fracture zones for as much as 13 km (8 mi). Most of these vein- and fracture-zone deposits are distributed within a 57 km^2 (22 mi^2) tract of Precambrian gneiss and migmatite located south and southeast of a quartz syenite complex at Democrat Creek. Christman and others (1953, 1959) mapped nearly 400 veins in this area. The dominant minerals forming these veins are smoky and clear quartz, microcline, barite, iron oxides, carbonates, and accessory rutile and sulfide minerals. Waxy, red thorite is the primary thorium mineral.

Thorium-REE minerals in the Wet Mountains district are also deposited in carbonatite dikes and small plugs. The carbonatite dikes are especially associated with the McClure Mountain complex (Staatz and Conklin, 1966). The carbonatites take a variety of forms, such as composite dikes with two or more generations of carbonate side by side with lamprophyre (Heinrich and Salotti, 1975; Armbrustmacher and others, 1979); phreatic explosion breccia pipes satellite to the McClure Mountain complex (the Pinon Peak breccias pipes of Heinrich and Dahlem, 1967); and siliceous carbonate dikes associated with amethyst veining (the Amethyst carbonatites of Heinrich and Shappirio, 1966). Armbrustmacher (1979) separated the carbonatites into two groups: replacement carbonatites and primary magmatic carbonatites. Replacement carbonatites have microscopic textures that indicate the nearly complete pseudomorphous replacement of relict igneous dike minerals by carbonate minerals. The replacement carbonatite dikes have ThO_2 contents of <0. 1 percent (Armbrustmacher and Brownfield, 1978). In contrast, the primary magmatic carbonatite dikes do not display mineral replacement textures and are enriched in elements and minerals typical of magmatic carbonatites, such as thorium, niobium, and REE that reside in the minerals thorite, bastnasite, synchysite, ancylite, and monazite. Thorium concentrations in the primary magmatic carbonatite dikes commonly exceed 0.1 percent ThO_2. However, thorium is more concentrated and is present in greater volume in the quartz–iron oxide–barite vein and fracture-zone deposits of the district in comparison with the carbonatites (Armbrustmacher, 1988).

From 52 samples of primary magmatic carbonatite in the Wet Mountains area, Armbrustmacher (1988) found average concentrations of 0.17 percent ThO_2 and 2.15 percent total rare earth oxides. Armbrustmacher (1988) calculated that the seven largest carbonatite dikes in the district contain the following:

- ThO_2
  Reserves of 119 metric tons (131 tons);
  Probable potential resources of 683 metric tons (753 tons);
- Total REE oxides
  Reserves of 2,270 metric tons (2,500 tons);
  Probable potential resources of 12,970 metric tons (14,300 tons).

Figure 11. Sewell Ranch thorium vein (between yellow lines), Wet Mountains, Custer County, south-central Colorado. This northwest-southeast-trending Cambrian vein, 9 ft (2.7 m) wide here, cuts about perpendicular to foliation of Precambrian mafic gneiss country rock. Analysis of outcrop samples of this vein collected by this study found 762 ppm Th (0.087 percent Th oxide); 495 ppm La; 1,280 ppm Ce; 752 ppm Nd; 296 ppm Sm; 78.5 ppm Eu; 174 ppm Gd; 16.2 ppm Tb; 14.5 ppm Ho; 4.16 ppm Tm; 25.2 ppm Yb; and 3.42 ppm Lu.

The Wet Mountains area also contains thin ($\leq$2 m thick) red syenite dikes that contain anomalous thorium and REE, particularly where the dikes are located nearest the intrusive centers. The syenite dikes are composed primarily of alkali feldspar and ferric oxides, with trace amounts of thorite, barite, rutile, xenotime, bastnasite, and brockite (Armbrustmacher, 1988). Samples of red syenite veins analyzed by this study found only 30–40 parts per million (ppm) Th, equivalent to 0.0034–0.0046 percent ThO_2, and 590–680 ppm total rare earth elements. Cerium accounts for about 40 percent of the rare earth content of the syenite dikes.

Sampling and geochemical analyses completed during this study showed that the syenite, mafic, and ultramafic rock units that form the core of the three intrusive complexes do not contain particularly large concentrations of thorium or REE. Samples of the quartz syenite pluton of the complex at Democrat Creek had an average content of 62 ppm Th (0.007 percent ThO_2) and 700 ppm total rare earth elements. Gabbro and pyroxenite units of the Gem Park complex showed no greater than 13 ppm Th and average total rare earth elements content of 190 ppm. In the McClure Mountain complex, all samples of the plutonic units of hornblende-biotite syenite (figure 12), nepheline syenite, pyroxenite, and gabbro contained less than10 ppm Th and no more than 355 ppm total rare earth elements content.

Figure 12. West-facing view of McClure Mountain, Fremont County, Colorado. The mountain is composed of a hornblende-biotite syenite phase of McClure Mountain complex. This stock is genetically related to thorium-rare earth elements vein deposits of Wet Mountains area. However, this rock unit and other core units of three intrusive complexes in this area contain only modest concentrations of thorium and rare earth elements. The thorium- and rare earth–bearing minerals crystallized in epigenetic vein and fracture-zone deposits distal to the alkaline intrusive complexes.

IDAHO—DIAMOND CREEK AREA

Location: This vein district lies on the eastern slope of the Salmon River Mountains, about 13 km (8 mi) north- northwest of Salmon, Idaho. Veins of the Diamond Creek district are found throughout an area only 4 km (2.5 mi) long by 0.8 km (0.5 mi) wide. Latitude: 45.29112 N., Longitude: 113.95174 W.; datum: WGS84

Deposit type and basic geology: The veins are hosted by Proterozoic quartzite and siltite and by Mesoproterozoic granite. The Diamond Creek veins are mineral fillings in fractured and sheared bedrock; the veins are as much as 7.6 m (25 ft) thick in the metasedimentary rocks (quartzite and siltite) but rarely more than 0.6 m (2 ft) thick in the granite (Staatz and others, 1979). These veins contain considerable amounts of hydrous iron oxide minerals, accompanied by disseminated thorium–rare earth elements–bearing minerals. The vein deposits with copious amounts of yellow-to-brown iron oxides (limonite and goethite) appear to contain the highest thorium (Th) and rare earth elements (REE) concentrations.

Status: No apparent exploration activity is underway in this district at present (2010). Within the last decade, Thorium Energy Inc. sampled vein deposits in the district.

Production: No thorium or rare earth elements have been produced from the district. In the Diamond Creek area, the larger veins were explored to evaluate their radioactivity in the latest 1940s into the early 1950s by use of bulldozed trenches, short adits, and hand-dug pits (Anderson, 1958). Most of the Diamond Creek area is soil covered, and thus the veins are exposed only by workings and road cuts.

Estimated resources: Sampling of the veins of the Diamond Creek district by the U.S. Geological Survey (Staatz and others, 1979, eight samples) found total REE oxide contents of 0.59 to 5.51 percent and thorium oxide contents of 0.04 to 1.71 percent (only one sample had more than 1 percent ThO_2). Most of the samples were more greatly enriched in the light REE compared with the heavy REE. Staatz and others (1979) estimated total reserves for the district of 2,870 tons (2,600 metric tons) of total rare-earth oxides and total probable potential resources of 75,500 tons (68,500 metric tons) of total REE oxides, using an average grade of 1.22 percent total REE oxides. Recent exploration and sample analysis by Thorium Energy Inc. supports the previous data reported by the USGS and the Atomic Energy Commission., including average total REE content of 0.80 percent and thorium content of 0.12 percent.

Detailed Discussion

The Diamond Creek district in eastern Idaho contains thorium- and rare earth elements (REE)-bearing veins similar to those in the Lemhi Pass district, 56 km (35 mi) to the southeast. This vein district lies on the eastern slope of the Salmon River Mountains, about 13 km (8 mi) north-northwest of Salmon, Idaho. Veins of the Diamond Creek district are found across an area of only 4 km (2.5 mi) long by 0.8 km (0.5 mi) wide (Anderson, 1958; Staatz and others, 1979). The veins are hosted by Proterozoic quartzite and siltite and by Mesoproterozoic granite (Evans and Zartman, 1990; Evans and Green, 2003). Similar in general appearance to the veins of Lemhi Pass, the Diamond Creek veins are mineral fillings in fractured and sheared bedrock; the veins are as much as 7.6 m (25 ft) thick in the metasedimentary rocks (quartzite and siltite), but rarely more than 0.6 m (2 ft) thick in the granite (Staatz and others, 1979).

In the Diamond Creek area, the larger veins were explored in the latest 1 940s into the early 1 950s by use of bulldozed trenches, short adits, and hand-dug pits (Anderson, 1958). Most of the Diamond Creek area is soil covered, and thus the veins are exposed only by workings and road cuts. Only eight veins have been identified in the district, traced along strike for 33.5 to 780 m (110 to 2,560 ft) (Anderson, 1958; Staatz and others, 1979). The eight veins vary from 0.15 to 7.6 m (0.5 to 25 ft) in thickness (Staatz and others, 1979). They contain considerable amounts of hydrous iron oxide minerals, accompanied by disseminated thorium-REE minerals.

In the Diamond Creek district, the vein deposits with copious amounts of yellow to brown iron oxides (limonite and goethite) appear to contain the highest thorium and REE concentrations (Anderson, 1958). Quartz, limonite, and goethite form the bulk of the veins, with locally abundant fluorite, potassium feldspar, hematite, and biotite (Anderson, 1958; Staatz and others, 1979). The primary thorium and REE mineral is monazite (Staatz and others, 1979). Thorium also occurs in minor amounts in brockite and thorite, and xenotime and bastnasite have been identified in trace amounts (Staatz and others, 1979).

Sampling of the veins of the Diamond Creek district by Staatz and others (1979, eight samples) found total REE oxide contents of 0.59 to 5.51 percent and thorium oxide contents of 0.04 to 1.71 percent (only one sample had more than 1 percent ThO_2). Most of the samples were more greatly enriched in the light REE than in the heavy REE. Staatz and others (1979) estimated total reserves for the district of 2,870 tons (2,600 metric tons) of total rare-earth oxides and total probable potential resources of 75,500 tons (68,500 metric tons) of total REE oxides, using an average grade of 1.22 percent total REE oxides.

Recent exploration and sample analysis by Thorium Energy Inc. supports the previous data reported by the USGS and the Atomic Energy Commission., including average total REE content of 0.80 percent and thorium content of 0.12 percent. Hedrick (2010) reports, "A preliminary estimate of the amounts of REE in the deposit are 22,400 t (metric tons) of cerium, 17,125 of neodymium, 8,220 t of lanthanum, 5,480 t of samarium, 4,795 t of yttrium, 3,425 t of praseodymium, 3,425 t of gadolinium, 1,370 t of dysprosium and europium each, and a total of 685 t of the remaining heavy rare earths." (One ton (2,000 lb) is equal to 0.9072 metric tons.)

IDAHO—HALL MOUNTAIN

Location: Veins crop out in an area of about 2.6 km^2 (1/2 mi^2)— 1,830 m (6,000 ft) by 305 m (1,000 ft) wide—on Hall Mountain in northernmost Idaho. Hall Mountain lies 1.6 km (1 mi) south of the United States–Canada border and 4.8 km (3 mi) east of the border station of Porthill, Idaho. Latitude: 48.98584 N., Longitude: 116.41887 W.; datum: WGS84

Deposit type and basic geology: The veins of Hall Mountain cut Precambrian quartzite and quartz diorite. They range in exposed length from 1.8 m (6 ft) to 213 m (700 ft) and in width from 0.18 m (0.6 ft) to 4 m (13 ft). Thorite is the primary thorium- and rare earth elements–bearing mineral; quartz and calcite are the most abundant gangue minerals, associated with chlorite magnetite, limonite, pyrite, and biotite, along with numerous minor and trace minerals. A total of 30 minerals were identified by Staatz (1972).

Status: No active exploration has been reported in this district.

Production: No mineral resources have been produced from this district.

Estimated resources: The U.S. Geological Survey (Staatz and others, 1979) determined that the possible reserves in this district are limited to thorium resources in only a few large veins. They estimate that the thorium reserves are 104,300 metric tons (115,000 tons) of vein material averaging 4.0

percent thorium oxide. They suggest that the rare earth elements are probably not economical in this district because of their low overall concentrations (average about 0.05 percent rare earth elements oxides).

Detailed Discussion

Thorium and REE-rich veins crop out in an area of about 2.6 km^2 (1/2 mi^2)—1,830 m (6,000 ft) by 305 m wide (1,000 ft)—on Hall Mountain in northernmost Idaho. Hall Mountain lies 1.6 km (1 mi) south of the United States-Canada border and 4.8 km (3 mi) east of the border station of Porthill, Idaho. Veins in this area can contain considerable thorium content, locally with as much as 21 percent ThO_2 (Staatz, 1972); however, their rare earth elements content is usually much less than their thorium content. As reported by Staatz and others (1974, p. 677), "Total rare-earth content of these veins ranges from 0.00111 to 0.197 percent in 12 samples from 10 veins; the thoria (ThO_2) content, from 0.011 to 5.84 percent." Staatz (1972, p. 240) reported, "The thorium content of 23 samples from 11 veins ranged from 0.0095 to 21 percent. Twelve samples had a thorium content greater than 1 percent."

The veins of Hall Mountain cut Precambrian quartzite and quartz diorite. They range in exposed length from 1.8 to 213 m (6 to 700 ft) and vary in width from 0.18 to 4 m (0.6 to 13 ft). Thorite is the primary thorium- and rare earth elements–bearing mineral; quartz and calcite are the most abundant gangue minerals, associated with chlorite magnetite, limonite, pyrite, and biotite, along with numerous minor and trace minerals. A total of 30 minerals were identified by Staatz (1972).

According to the analyses of Staatz and others (1979), the possible reserves in this district are limited to thorium resources in only a few large veins. They estimate that the thorium reserves are 104,300 metric tons (115,000 tons) of vein material averaging 4.0 percent ThO_2. They suggest that the rare earths are probably not economical in this district because of their low overall concentrations (average about 0.05 percent rare earth elements oxides). The distribution of rare earth elements in the Hall Mountain veins, on the basis of the average concentrations of 10 vein samples reported by Staatz and others (1974), is summarized in table 18.

Table 18. Mean Concentrations of Rare Earth Elements Measured in 10 Vein Samples, Hall Mountain, Idaho

[Rare earth elements listed in order of increasing atomic number; yttrium (Y) is included with these elements because it shares chemical and physical similarities with the lanthanides. wt percent, weight percent. Data averaged from analyses reported by Staatz and others (1974, table 1)]

Oxide	Average wt percent
La_2O_3	0.0013
Ce_2O_3	0.0053
Pr_2O_3	<0.0006
Nd_2O_3	0.0016
Sm_2O_3	0.0011
Eu_2O_3	<0.0018
Gd_2O_3	0.0041
Tb_2O_3	<0.0018
Dy_2O_3	0.0034
Ho_2O_3	<0.0012
Er_2O_3	0. 0024
Tm_2O_3	<0.0017
Yb_2O_3	0.0025
Lu_2O_3	<0. 0014
Y_2O_3	0.03 15
Total	0.0512

IDAHO—LEMHI PASS DISTRICT, IDAHO- MONTANA

Location: The Lemhi Pass district contains numerous vein deposits enriched in thorium and rare earth elements (REE) within a 140 km^2 (54 square mi) core of a larger 400 km^2 (154 mi^2) area in the central Beaverhead Mountains; the district straddles the Continental Divide on the Montana-Idaho border. Latitude: 44.93728 N., Longitude: 113.46451 W.; datum: WGS84

Deposit type and basic geology: Within the Lemhi Pass district, Staatz (1972; Staatz and others, 1979) mapped 219 veins enriched in thorium and rare earth elements (REE). Most of these veins are quartz-hematite-thorite veins that fill fractures, shears, and brecciated zones in Mesoproterozoic quartzite and siltite host rocks. Thorium and REE also appear in monazite-thorite-apatite shears and in replacements with specularite, biotite, and alkali feldspar. The thorium-REE veins of the district range from 1 m (3.3 ft) to at least 1,325 m (4,347 ft) in length and from a few centimeters (1 in.) to as much as 12 m (39 ft) in width. The Last Chance vein—1,325 m (4,348 ft) long and 3–8 m (10–26 ft) wide for most of its length—is the longest and widest vein in the district.

Status: During the last decade (prior to 2010), Thorium Energy Inc. has evaluated the thorium and REE resources in the vein systems of the district.

Production: No thorium or REE have been produced from this district. Past exploration and development in these vein deposits focused on their thorium content; earlier development trenched several veins and produced modest underground workings in the Last Chance vein.

Estimated resources: This district is thought to represent the largest concentration of thorium resources in the United States (Van Gosen and others, 2009). On average, the thorium veins of the district have roughly equal concentrations of thorium and total rare earth elements. Thus, the REE resources of the vein deposits of the Lemhi Pass district are approximately equal to its thorium resource. Earlier studies by the USGS estimated that the Lemhi Pass district contains total reserves of 64,000 metric tons (70,500 tons) of thorium oxide (ThO_2) and probable potential resources of an additional 121,000 metric tons (133,000 tons) (Staatz and others, 1979). The 10 largest veins, with an average grade of 0.43 percent ThO_2, represent 95 percent of the district's identified thorium resources. Using a compilation of surface, underground, and drilling assays, the Idaho Energy Resource Company reported a "quantitative proven" reserve of 176 metric tons (194 tons) of ThO_2 within the Last Chance vein and a possible resource of 2,000 metric tons (2,200 tons) of additional ThO_2 (Idaho Energy Resource Company, written commun., 2008).

Detailed Discussion

The Lemhi Pass district contains numerous vein deposits enriched in thorium and rare earth elements (REE) within a 140 km^2 (54 mi^2) core of a larger 400 km^2 (154 mi^2) area in the central Beaverhead Mountains; the district straddles the Continental Divide on the Montana-Idaho border (figure 13).

This district is thought to represent the largest concentration of thorium resources in the United States (Van Gosen and others, 2009). Earlier studies by the USGS estimated that the Lemhi Pass district contains total reserves of 64,000 metric tons (70,550 tons) of thorium oxide (ThO_2) and probable potential resources of an additional 121,000 metric tons (133.400 tons) (Staatz and others, 1979). The 10 largest veins, with an average grade of 0.43 percent ThO_2, represent 95 percent of the district's identified thorium resources. Using a compilation of surface, underground, and drilling assays, the Idaho Energy Resource Company reported a "quantitative proven" reserve of 176 metric tons (194 tons) of ThO_2 within the Last Chance vein and a possible resource of 2,000 metric tons (2,200) of additional ThO_2 (Idaho Energy Resource Company, written commun., 2008). On average, the thorium veins of the district have roughly equal concentrations of thorium and total rare earth elements. Thus, the REE resources of the vein deposits of the Lemhi Pass district are approximately equal to its thorium resource. The Last Chance vein and the Wonder vein (figure 14) are the only deposits in the district that have been sampled by underground or drill-hole access. Much exploration potential exists in the district.

Within the Lemhi Pass district, Staatz (1972) and Staatz and others (1979) mapped 219 veins enriched in thorium and REE. Most of these veins are quartz-hematite-thorite veins, which fill fractures, shears, and brecciated zones in Mesoproterozoic quartzite and siltite host rocks. Thorium and REE also are present in monazite-thorite-apatite shears and replacements with specularite, biotite, and alkali feldspar. The thorium-REE veins of the district range from 1 m (3.3 ft) to at least 1,325 m (4,347 ft) in length and from a few centimeters (1 in.) to as much as 12 m (39 ft) in width. The Last Chance vein—1,325 m (4,348 ft) long and 3–8 m (10–26 ft) wide for most of its length—is the longest and widest vein in the district; this vein also represents the largest individual thorium and REE resource in the district. Fifteen thorium veins in the district exceed 300 m (984 ft) in length. Some of the veins contain carbonate minerals, such as calcite, siderite, and ankerite, and local fluorite. Rare earth elements- and thorium-bearing allanite and monazite are locally abundant. Other reported ore minerals include brockite, xenotime, and thorite. The primary gangue minerals are quartz, hematite, limonite, apatite, potassium feldspar, biotite, albite, and barite. Most of the veins are extensively weathered and have abundant iron-oxide staining. The district also hosts small quartz-copper-gold (and rare molybdenum) veins, and some of the thorium veins contain very small amounts of base metals, such as copper, iron, manganese, lead, and zinc.

Figure 13. View to west of Lemhi Pass, Idaho-Montana. The ridge, a part of Beaverhead Mountains, forms a segment of the Continental Divide and the Idaho-Montana border; Montana is in foreground and Idaho is in the distance. More than 200 rare earth elements- and thorium-rich veins in this area form the Lemhi Pass district.

Figure 14. Outcrop of Wonder vein (between red lines), Lemhi Pass district, Idaho-Montana, exposed in a mined bench. Vein is heavily oxidized and consists mainly of silica, likely some carbonate, and iron oxide minerals with thorite and altered thorite. Host rock is Precambrian quartzite and siltite.

Table 19. Mean Concentrations of Rare Earth Elements Measured in Samples of Last Chance Vein, Idaho-Montana

[Rare earth elements listed in order of increasing atomic number; yttrium (Y) is included with these elements because it shares chemical and physical similarities with the lanthanides. wt percent, weight percent. Last Chance is the largest vein in the Lemhi Pass district, Idaho-Montana. Data averaged from nine samples of the vein reported by Staatz (1972, table 5)]

Oxide	Average wt percent
La_2O_3	0.033
Ce_2O_3	0.082
Pr_2O_3	0.014
Nd_2O_3	0.127
Sm_2O_3	0.087
Eu_2O_3	0.027
Gd_2O_3	0.056
Tb_2O_3	0.003
Dy_2O_3	0.008
Ho_2O_3	<0.003
Er_2O_3	0.002
Tm_2O_3	<0.003
Yb_2O_3	<0.003
Lu_2O_3	<0.003
Y_2O_3	0.015
Total	0.454

The thorite veins of the Lemhi Pass district are approximately equally enriched in thorium and REE; the ratio of thorium to REE concentrations in the veins averages around 1:1. Staatz (1972) reported the REE analyses of 31 vein samples, which showed total REE-oxide contents ranging from 0.073 to 2.20 percent, with an average value of 0.428 percent (very similar to the average thorium oxide content of 0.43 percent found in the 10 largest veins in the district). Table 19 lists the average distribution of the rare earth elements reported by Staatz (1972, table 5) from the analyses of nine samples of the Last Chance vein. The district's thorium veins are most commonly enriched in the middle REE (especially neodymium), with some veins apparently enriched more in the heavy REE (Staatz, 1972, p. 76–77). Using modern techniques in recent analytical work, the Idaho Geological Survey and industry (Idaho Energy Resource Company in 1991; Thorium Energy in 2008) has confirmed the overall thorium and REE concentrations and the unusual enrichments in middle REE-group minerals. Hedrick (2010) reported, “Based on average percentages of individual REE by recent sampling and previous analyses by Idaho Energy Reserves Co. (a subsidiary of Idaho Power Co.), the Lemhi Pass District had resources, in order of increasing atomic number, 77,345 t of yttrium; 25,780 t of lanthanum; 69,980 t of cerium; 11,050 t of praseodymium; 66,296 t of neodymium; 40,515 t of samarium; 14,735 t of europium; 40,515 t of gadolinium; 1,840 of terbium; 14,730 t of dysprosium; 1,840 t each of holmium and ytterbium; and about 929 t or less each of erbium, thulium, and lutetium.”

Illinois—Hicks Dome

Location: Hicks Dome is located in Hardin County, southernmost Illinois. Latitude: 37.53131 N., Longitude: 88.36873 W.; datum: WGS84

Deposit type and basic geology: The dome-shaped structure, approximately 14.5 km (9 mi) in diameter, was formed by the displacement of sedimentary rocks at least 1,200 m (3,940 ft) upward above an alkaline intrusion at depth. More than 600 m (1,970 ft) of sedimentary rocks, mostly limestone, were pushed up by the explosive intrusion of magmatic fluids. A hole drilled near the apex of the dome (Brown and others, 1954) intersected a mineralized breccia at a depth of 490 m (1,607 ft), which continues to the bottom of the hole at 897 m (2,944 ft). Mineralization in the breccia contains thorium and rare earth elements, tentatively identified as residing in monazite, and is found in association with florencite, a cerium-aluminum phosphate; gangue minerals are fluorspar, calcite, quartz, minor pyrite, and traces of sphalerite and galena.

Status: Currently (2010), no exploration appears to be active at this feature.

Production: No mineral resources have been produced from this intrusive complex.

Estimated resources: Resources have not been estimated. Eight samples of drill core, each 7.6–9.1 m (25–30 ft) long, contained thorium concentrations of 0.007–0.18 percent thorium oxide (Brown and others, 1954). Rare earth elements content in the eight drill-core samples was 0.1–0.99 weight percent yttrium; 0.01–0.099 weight percent lanthanum, cerium, neodymium, and dysprosium; and 0.001–0.0099 weight percent ytterbium (Trace, 1960). These samples represent only 64 m (210 ft) of a breccia zone enriched in thorium and rare earth elements; this zone could extend throughout a large area across the roof of Hicks Dome.

Detailed Discussion

Hicks Dome, in Hardin County of southernmost Illinois, overlies a potentially wide area of rare earth elements (REE) and thorium (Th) mineralization at depth. The dome- shaped structure, approximately 14.5 km (9 mi) in diameter, was formed by the displacement of sedimentary rocks at least 1,200 m (3,940 ft) upward above an alkaline intrusion at depth. More than 600 m (1,970 ft) of sedimentary rocks, mostly limestone, were pushed up by the explosive intrusion of magmatic fluids (Heyl and others, 1965). A hole drilled near the apex of the dome (Brown and others, 1954) intersected a mineralized breccia at a depth of 490 m (1,607 ft), which continues to the bottom of the hole at 897 m (2,944 ft). Mineralization in the breccia includes fluorspar, calcite, quartz, minor pyrite, and traces of sphalerite and galena. Eight samples of this drill core, each 7.6–9.1 m (25–30 ft) long, contained 0.007 to 0.18 percent ThO_2 (Brown and others, 1954). REE content in the eight drill-core samples was 0.1–0.99 weight percent Y; 0.01–0.099 weight percent for La, Ce, Nd, and dysprosium (Dy); and 0.001–0.0099 weight percent ytterbium (Yb) (Trace, 1960). These samples represent only 64 m (210 ft) of a breccia zone enriched in Th-REE; this zone could extend throughout a large area across the roof of Hicks Dome.

Shallow diamond drilling and trenching sampled an area of radioactive breccia atop Hicks Dome in which the radioactive mineral was tentatively identified as monazite, and found, in association with florencite, a cerium-aluminum phosphate (Trace, 1960). A surface sample from a trench contained 0.1–0.5 weight percent Th plus REE, including 0.5–1 weight percent Ce and La, 0.1–0.5 weight percent Nd, 0.05–0.1 weight percent praseodymium (Pr), 0.01–0.05 weight percent terbium (Tb), and 0.005–0.01 weight percent Yb (Trace, 1960).

Using airborne gamma-ray data, Pitkin (1974) delineated the large extent of the radioactivity anomaly at Hicks Dome, which arises because of its thorium content. However, the apparent depth of this REE-thorium deposit may limit its resource potential. Much more surface and subsurface exploration is necessary to evaluate the extent and grade of this deposit.

MISSOURI—PEA RIDGE IRON DEPOSIT AND MINE

Location: The Pea Ridge iron orebody and mine site is located in Washington County, Missouri, about 97 km (60 mi) southwest of St. Louis. Latitude: 38.12621 N., Longitude: 91.04766 W.; datum: WGS84

Deposit type and basic geology: Rare earth elements (REE)–bearing breccia pipes cut through the Pea Ridge massive magnetite-iron orebody. The Pea Ridge deposit is hosted by Precambrian volcanic rocks of the St. Francois terrane of southeastern Missouri. The magnetite-rich orebody is interpreted as a high-temperature, magmatichydrothermal deposit (Sidder and others, 1993) in ash-flow tuffs and lavas, which may have formed in the root of a volcanic caldera (Nuelle and others, 1991). Four mapped REE-bearing breccia pipes steeply crosscut the magnetite-hematite orebody and its altered rhyolite host rock. Exposed portions of the breccia pipes are as much as 60 m (197 ft) in horizontal length and as much as 15 m (49 ft) in width; the pipes extend below the mined levels to an undetermined depth (Seeger and others, 2001). Rare earth elements–bearing minerals in the breccia pipes include monazite, xenotime, and minor bastnasite and britholite. The REE concentrations reported in the breccia pipes are consistently high but variable. Nuelle and others (1992, p. A1) state, "Total REE oxide content of samples of the groundmass material, which are not diluted with lithic fragments, average about 20 weight percent." Seeger and others (2001, p. 2) state, "Total REE oxide concentrations of grab samples range from about 2.5 to 19 weight percent." Bulk sampling by the U.S. Bureau of Mines found REE oxides concentrations ranging from 7 to 25 weight percent and an average of 12 weight percent (Vierrether and Cornell, 1993).

Status: Currently (2010), there is no active development in this deposit. In 2005, Upland Wings formed Wings Enterprises, with the intent to reclaim iron ore at the site and produce iron from its large surface reserves (http://www.wingsironore.com/).

Production: In 1957, the deposit was first developed by the Bethlehem Steel and St. Joseph Lead Company, under the name Meramec Mining

Company. In 1964, iron production began from this deposit. The mine operated from 1964 to 2001 with three interruptions, producing more than 30 million tons (27 million metric tons) of pellets, fines, heavy media, and other iron products. In 1990, the mine lost its last iron-ore-pellet customer and began to produce specialty products. The Pea Ridge mine continued to operate while iron ore prices fell during the 1 990s, but in 2001 the mine went into bankruptcy. In 2001, Upland Wings, Inc., purchased the Pea Ridge Iron Ore mine properties and all of its mineral rights.

Estimated resources: A U.S. Bureau of Mines report by Whitten and Yancey (1990) estimated that the breccia pipes contain about 600,000 metric tons (660,000 tons) of REE reserves with an average grade of 12 percent REE oxides. The report does not indicate the data used to calculate this estimate. However, a similar value of 600,000 short tons is mentioned as a note on another internal company memo (provided by Jim Kennedy, electronic communication, October 2008); that memo is dated 10-25-89 and signed by Larry J. Tucker (retired Pea Ridge mine superintendent); supporting calculations for this value are missing. A copy of another internal company memo (provided by Jim Kennedy, electronic communication, October 2008 and dated 11-22-88 and signed by Larry J. Tucker) indicates that there is a combined, probable reserve in two of the breccia pipes of approximately 250,000 metric tons (276,000 tons) of mineralized rock, grading about 13 percent REE (note: weight percent, not oxide equivalent). The surface tailings contain additional lanthanide resources primarily in fine-grained, REEbearing minerals, chiefly monazite and xenotime, that form inclusions within apatite. The apatite also contains minor amounts of REE in its structure; apatite is found in variable concentrations throughout the iron orebody (Vierrether and Cornell, 1993).

Detailed Discussion

Rare earth elements (REE)–bearing breccia pipes cut through the Pea Ridge massive magnetite iron-orebody in Washington County, Missouri, about 97 km (60 mi) southwest of St. Louis. The iron deposit as a whole contains concentrations of REE that may be economically recoverable as a primary product or as a byproduct of iron ore production.

The Pea Ridge massive magnetite deposit is hosted by Precambrian volcanic rocks of the St. Francois terrane of southeastern Missouri; this volcanic-plutonic province is composed of Mesoproterozoic rhyolitic ash-flow tuffs, lava flows, and granitic plutons (Kisvarsanyi, 1980). The St. Francois terrane contains eight known Mesoproterozoic magnetite-hematite deposits and forms an iron metallogenic province (Kisvarsanyi and Proctor, 1967) that hosts nearly 1 billion metric tons of identified ore (Arundale and Martin, 1970). The Pea Ridge massive magnetite orebody has been interpreted as a high- temperature, magmatic-hydrothermal deposit (Sidder and others, 1993) in ash-flow tuffs and lavas, which may have formed in the root of a volcanic caldera (Nuelle and others, 1991).

The Pea Ridge deposit is covered by Cambrian and Ordovician sedimentary rocks that unconformably overlie the deposit and the underlying Precambrian rocks. The orebody lies discordant to the host volcanic rocks, striking N. 60° E. with a nearly vertical dip, whereas the host volcanic rocks strike N. 80° W. and dip 75° NE. (Emery, 1968). The primary host for the iron orebody is altered rhyolite tuff (Nuelle and others, 1992). This massive iron deposit is estimated to contain more than 100 million short tons of ore (Arndt, 1981).

The deposit area contains nine mappable rock units: amphibole-quartz rock; heterolithic breccia; pseudobreccia; magnetite; hematite; silicified rock; REE mineral-bearing breccia pipes; mafic dikes; and aplite dikes (Nuelle and others, 1992; Seeger and others, 2001). Four mapped lanthanide-bearing breccia pipes (X11, V12, X13, V14) steeply crosscut the magnetite-hematite orebody and its altered rhyolite host rock (figure 15); the pipes are situated along the footwall and eastern edge of the iron orebody (Seeger and others, 2001). Exposed portions of the breccia pipes are as much as 60 m (197 ft) in horizontal length and as much as 15 m (49 ft) in width; the pipes extend below the mined levels to an undetermined depth (Seeger and others, 2001). One of the pipes is exposed for 120 m (394 ft) vertically. As described by Seeger and others (2001, p. 2), the four breccia pipes of the Pea Ridge deposit consist of the following:

> "Fragments of rhyolite, iron oxide, and silicified rock in a groundmass of rock flour, feldspar, chlorite, barite, apatite, monazite, xenotime, quartz, and calcite. Volcanic rock fragments range from less than 1 mm to about 0.5 m in diameter, are subrounded to angular with moderate to high sphericity, and have undergone potassium metasomatism. Specularite [iron oxide] fragments are angular, are as long as several meters, and have low to moderate sphericity."

The REE-bearing minerals in the breccia pipes include monazite, xenotime, and minor bastnasite and britholite. Nuelle and others (1992) describe the monazite and xenotime as forming radial crystal aggregates and granular crystals 0.5–1.9 mm (0.02—0.75 in.) across; these minerals also replace microfragments in the wall rock and within the groundmass fill fractures in barite and potassium feldspar crystals. The mineralogy of Pea Ridge is summarized by Nuelle (1998), and its mineral paragenesis and alteration zones are described by Sidder and others (1993). Gold, tin, and silver are unevenly distributed in the breccia pipes and in both the hematite and silicified zones (Husman, 1989). Nuelle and others (1992) report localized gold concentrations as much as 371 parts per million.

The REE oxide concentrations are relatively high within the breccia pipes of the Pea Ridge deposit. The REE abundances reported in the breccia pipes are consistently high but variable. Nuelle and others (1992, p. A1) state, "Total REE oxide content of samples of the groundmass material, which are not diluted with lithic fragments, average about 20 weight percent." Seeger and others (2001, p. 2) state, "Total REE oxide concentrations of grab samples range from about 2.5 to 19 weight percent." Bulk sampling by the U.S. Bureau of Mines found REE oxides concentrations ranging from 7 to 25 weight percent and an average of 12 weight percent (Vierrether and Cornell, 1993). The size of the REE resource in the four breccia pipes has not been determined and these pipes are open downward. A copy of an internal company memo (provided by Jim Kennedy, electronic communication, October 2008, dated 11-22-88 and signed by Larry J. Tucker, retired Pea Ridge mine superintendent) indicates that there is a combined, probable reserve for pipes X11 and X13 of approximately 250,000 metric tons of mineralized rock, grading about 13 percent REE (note: weight percent, not oxide equivalent). The reserves were calculated for the volume between levels 2275 and 2675 by using a density of 2,000 lb per 9 cubic foot (ft^3) (approximately 3.56 gram per cubic centimeter (g/cm^3)). (Working levels in the mine, now flooded, are named for their depth below the collar of the shaft; the uppermost level was at 1,375 ft and the deepest level was at 2,675 ft.)

A U.S. Bureau of Mines report by Whitten and Yancey (1990) estimated that the breccia pipes contain about 600,000 metric tons of REE reserves with an average grade of 12 percent REE oxides. The report does not indicate the data used to calculate the 600,000 metric ton resource estimate. However, a similar value of 600,000 short tons is mentioned as a note on another internal company memo (provided by Jim Kennedy, electronic communication, October 2008) that is dated 10-25-89 and signed by Larry J. Tucker;

supporting calculations for this value are missing. The surface tailings contain additional lanthanide resources, primarily in fine- grained, REE-bearing minerals, chiefly monazite and xenotime, that form inclusions within apatite. The apatite also contains minor amounts of REE in its structure; apatite is found in variable concentrations throughout the iron orebody (Vierrether and Cornell, 1993).

The rare earth elements in the breccia pipes of the Pea Ridge are dominated by the light REE lanthanum (La) and cerium (Ce), but the pipes are also relatively enriched in heavy REE, including dysprosium (Dy), holmium (Ho), erbium (Er), ytterbium (Yb), lutetium (Lu), and yttrium (Y) (table 20). The REE (lanthanide) resources of Pea Ridge are proportionally more concentrated in these heavy REE than is true of most other U.S. deposits (table 21). The REE distribution in the Pea Ridge deposit was generally confirmed by a recent USGS study, which collected limited samples of tailings and drill core at the site (Grauch and others, 2010). Although the total REE oxide resource at Pea Ridge (72,000 metric tons) is very small in comparison with the Mountain Pass deposit (2.58 million metric tons), the Pea Ridge deposit may be viewed as a potential source of heavy REE as a byproduct of iron ore production.

The Pea Ridge magnetite deposit was identified in 1950 from a prominent magnetic anomaly. In 1957, the deposit was first developed by the Bethlehem Steel and St. Joseph Lead Company, under the name Meramec Mining Company. In 1964, production began from this deposit. The mine operated from 1964 to 2001, with three interruptions, and produced more than 30 million tons (27 million metric tons) of pellets, fines, heavy media, and other iron products. In 1990, the mine lost its last iron-ore-pellet customer and began to produce specialty products (information from the Wing Enterprises Web site, http://www.wingsironore.com/). In contrast, Nuelle (1998) reported a much higher historic production and stated, “to date [March 1998], Pea Ridge has produced 50.7 million metric tons of iron ore.”

The Pea Ridge mine continued to operate while iron ore prices fell during the 1990s, but in 2001 the mine went into bankruptcy. Later that year, Upland Wings, Inc. purchased the Pea Ridge Iron Ore mine properties and all of its mineral rights. In 2005, Upland Wings formed Wings Enterprises, with the intent to reclaim iron ore at the site and produce iron from its large surface reserves. Wings Enterprises suggests that its reclamation production facility can produce more than 30 short tons per hour of 70 percent Fe (95 percent magnetite) ore material from its estimated 300,000 short tons of surface reserve.

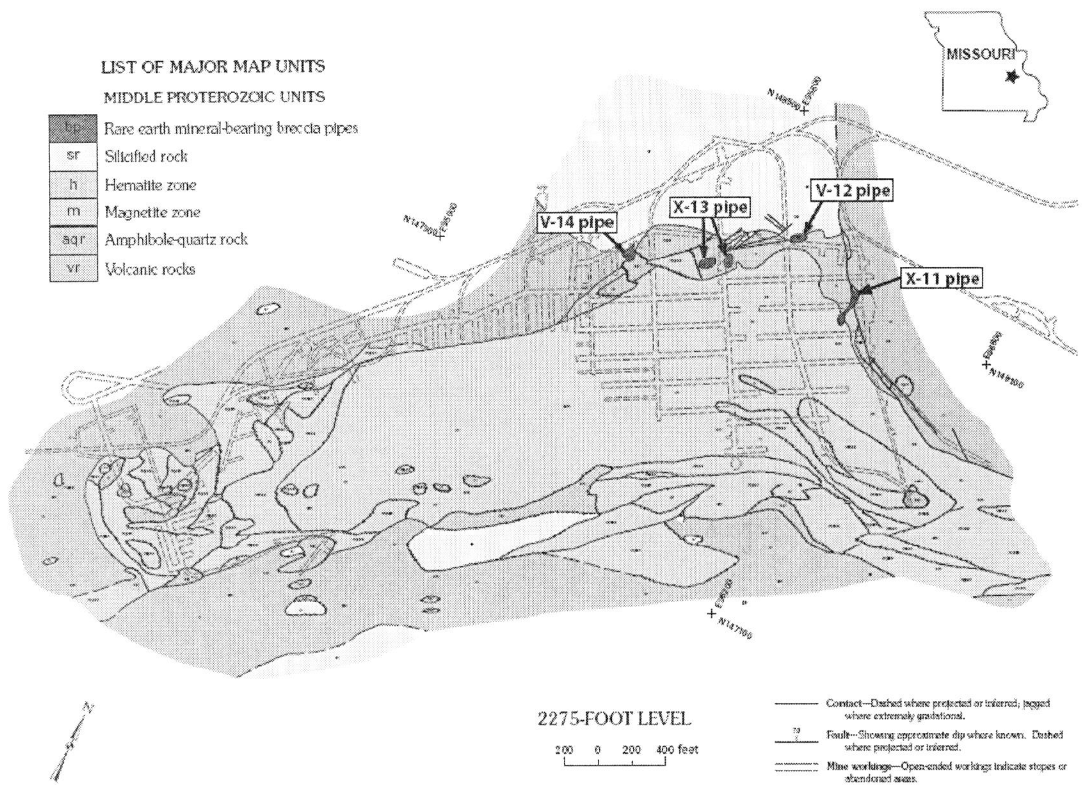

Figure 15. Generalized geologic map of the 2275 level of the Pea Ridge iron mine, Washington County, Missouri. Map from Grauch and others (2010), who adapted it from Seeger and others (2001).

Table 20. Rare Earth Elements and Thorium Concentrations in Four Breccia Pipes, Pea Ridge Deposit, Missouri

[Rare earth elements listed in order of increasing atomic number; yttrium (Y) is included with these elements because it shares chemical and physical similarities with the lanthanides. --, not available. Data from internal company memo by Larry J. Tucker, dated 10-25-89 (Jim Kennedy, electronic communication, October, 2008)]

Element	Breccia pipe			
	X-11 (percent)	V-12 (percent)	X-13 (percent)	V-14 (percent)
La	4.45	2.70	2.95	2.05
Ce	8.00	4.50	4.95	4.05
Pr	0.68	--	0.41	0.34
Nd	2.15	--	1.50	1.10
Sm	0.42	--	0.33	0.24
Eu	0.03	--	0.03	0.02
Gd	0.15	--	0.18	0.08
Tb	--	--	--	--
Dy	0.19	--	0.18	0.09
Ho	0.03	--	0.03	0.01
Er	0.09	--	0.09	0.04
Tm	--	--	--	--
Yb	0.16	--	0.12	0.06
Lu	0.02	--	0.01	--
Y	0.70	0.69	0.67	0.36
Th	0.63	--	0.23	0.41

The Pea Ridge mine site, currently inactive (2010), contains an iron mine, mill, iron-pellet-making facility, and large piles of milled iron-ore tailings resulting from its earlier operation. The property now has two large tailings lakes flanked by extensive waste and tailings piles; most have vegetation cover, and several contain smaller ponds and wetlands. There are also several small, dry tailings ponds and a variety of ore stockpiles. Tailings underlie approximately 180 acres in total. The entire property is currently owned by Jim Kennedy, the president of Wings Enterprises, Inc., in Saint Louis, Missouri (http://www. wingsironore.com/).

Table 21. Rare Earth Elements Oxide Concentrations of Two World-Class Chinese Rare Earth Elements Deposits Compared with Concentrations in Major United States Rare Earth Elements Deposits

[Rare earth elements listed in order of increasing atomic number; yttrium (Y) is included with these elements because it shares chemical and physical similarities with the lanthanides. --, not available. Values listed below were calculated by combining data from sources providing either tonnage or rare earth elements distribution]

Rare earth oxide	Bayan Obo deposit[1] (metric tons)	China Clay deposits[2] (metric tons)	Mountain Pass deposit[3] (metric tons)	Iron Hill carbonatite[4] (metric tons)	Pea Ridge deposit[5] (metric tons)	Lemhi Pass district[6] (metric tons)
La_2O_3	15,267,052	193,001	872,120	264,469	18,275	4,672
Ce_2O_3	23,720,328	311,762	1,279,918	522,907	32,298	11,584
Pr_2O_3	1,734,339	40,529	106,337	68,838	2,862	1,984
Nd_2O_3	6,083,552	135,197	288,040	257,716	9,474	17,903
Sm_2O_3	516,088	38,165	21,939	36,340	1,963	12,264
Eu_2O_3	99,822	3,430	2,710	8,352	158	3,776
Gd_2O_3	247,460	29,699	5,420	25,464	808	7,872
Tb_2O_3	23,884	4,282	413	2,806	--	448
Dy_2O_3	56,584	22,994	878	6,051	903	1,128
Ho_2O_3	12,594	4,918	103	826	137	--
Er_2O_3	14,270	14,401	155	1,973	430	256
Tm_2O_3	3,376	2,059	52	216	--	--
Yb_2O_3	2,364	10,755	52	1,195	662	--
Lu_2O_3	490	1,627	--	125	88	--
Y_2O_3	117,798	187,181	3,355	23,228	3,942	2,113
Total	48,000,000	1,000,000	2,581,490	1,220,506	72,000	64,000

[1] Bayan Obo deposit, Inner Mongolia, China: Berger and others (2009) indicate that Bayan Obo has 800 million metric tons of ore at 6 percent REE oxide content. Rare earth elements distribution is based on the average of three analyses of mineralized material from the East Ore Deposit; data from Yang and others (2009).

[2] China Clay deposits, southern China: Clark and Zheng (1991) indicate that the combined rare earth elements oxide content of all the China Clay deposits is at least 1 million metric tons. Rare earth elements distribution is based on a single ore concentrate (Grauch and others, 2010, table 4).

[3] Mountain Pass deposit, California: Castor and Nason (2004) indicate that Mountain Pass contains estimated reserves of 29 million metric tons of ore at 8.9 wt. percent rare earth elements oxides (by using a 5 percent cutoff). Rare earth elements distribution calculated from data in Castor (2008) from rare earth elements oxide contents in concentrate.

[4] Iron Hill carbonatite, Colorado: Staatz and others (1979) estimated that the carbonatite stock of Iron Hill consists of 655.6 million metric tons of carbonatite. Rare earth elements oxide resources were calculated from median concentrations measured by Van Gosen (2008) in 13 samples of the Iron Hill stock.

[5] Pea Ridge deposit, Missouri: Whitten and Yancey (1990) indicate that Pea Ridge contains 600,000 metric tons of ore with an average tenor of 12 percent rare earth elements oxides. Rare earth elements distribution is based on the average of composite assays of samples from four breccia pipes (Grauch and others, 2010).

[6] Lemhi Pass district, Idaho-Montana: Rare earth elements distribution is based on the average of analyses of nine samples of the Last Chance vein, reported by Staatz (1972).

Nebraska—Elk Creek Carbonatite

Location: Located near the small town of Elk Creek in southeastern Nebraska. Latitude: 40.26861 N., Longitude: 96.18333 W.; datum: WGS84

Deposit type and basic geology: A buried, rare earth elements (REE)- and niobium (Nb)-rich carbonatite mass, referred to as the Elk Creek carbonatite, lies in the subsurface about 1.6 km (1 mi) southwest of the small town of Elk Creek in southeastern Nebraska. On the basis of exploration drilling and the extent of magnetic and gravity anomalies, the carbonatite mass at depth appears to have its center beneath section 33, township 4 north, range 11 east. The entire oval-shaped, subsurface body, which is recognized by a geophysical anomaly caused by the carbonatite and associated intrusive rocks, is about 7 km (4.3 mi) in diameter. Analyses of drill core showed the intrusion at depth comprised mostly massive to brecciated, apatite- and pyrochlore-bearing dolomitic carbonatite (89 percent), along with fenitized basalt, lamprophyre, and syenite (totaling 11 percent). Major-element analyses suggest that the carbonate mass is a magnesian carbonatite (dolomitic), generally similar in gross chemical composition to the Iron Hill (Powderhorn) carbonatite stock in southwestern Colorado. The REE are hosted principally by the minerals bastnasite, parisite, and synchisite and by smaller amounts of monazite (Xu, 1996). Niobium was deposited in pyrochlore. The U.S. Geological Survey obtained a potassium-argon age on biotite in the carbonatite of 544±7 million years old (Xu, 1996).

Status: On May 4, 2010, Quantum Rare Earth Developments Corp. announced that it had acquired the Elk Creek carbonatite properties (http://www.quantumrareearth.com/).

Production: No mineral resources have been produced from this intrusion.

Estimated resources: It has been reported that the Elk Creek carbonatite may represent the largest niobium (Nb) resource in the United States. Quantum Rare Earth Developments Corporation reported several assay results from Molycorp's earlier drilling program. Reportedly, drilling within the core zone found high-grade niobium contents, estimated at "39.4 million tons of 0.82 percent Nb2O5 and is open to the north, west and at depth (Molycorp, Inc., internal memorandum, Feb 05/1 986)." In the widely spaced drilling surrounding the core zone, "at least 18 of the surrounding holes intersected greater than 6.1 m (20 ft) of greater than 1.0 percent REO (total rare earth oxides), while at least 17 of the surrounding holes intersected greater than 3.05 m (10 ft) of greater than 0.6 percent Nb_2O_5." Quantum reports assay intervals that range from 1.02 to 3.12 percent total rare earth elements oxide. They also note, "Most of the historic drill core, sample rejects, and pulps from Molycorp's exploration are available for review and sampling."

Detailed Discussion

A buried, rare earth elements (REE)– and niobium-rich carbonatite mass, referred to as the Elk Creek carbonatite, lies in the subsurface about 1.6 km (1 mi) southwest of the small town of Elk Creek in southeastern Nebraska. On the basis of exploration drilling results and the extent of magnetic and gravity anomalies, the carbonatite mass at depth appears to have its center beneath section 33, township 4 north, range 11 east. The entire oval-shaped, subsurface body, which is recognized by a geophysical anomaly caused by the carbonatite and associated intrusive rocks, is about 7 km (4.3 mi) in diameter; it straddles the boundary between Johnson County and Pawnee County (Carlson and Treves, 2005).

In 1970, a regional geophysical program detected a nearly circular, concurrent magnetic and gravity anomaly in this area. This area of Nebraska is blanketed by loess and glacial till that overlies Pennsylvanian marine carbonates and shale. Exploratory drilling in this area had previously encountered Precambrian granitic and metamorphic rocks at depths of 600 ft (183 m) (Carlson and Treves, 2005). Modeling by Burfeind and others (1971) of the geophysical data collected over the anomaly suggested a cylindrical body with an indefinite length and a radius of 5,500 ft (1,676 m), which was beveled on the basement surface at a depth of about 600 ft (183 m). A test hole was drilled into the anomaly, which found 45 ft (13.7 m) of unconsolidated cover of Quaternary loess and glacial till overlying 583 ft (178 m) of Upper and Middle Pennsylvanian carbonates and shale. At a depth of 630 ft (192 m), the drilling hit an iron-rich, silicate-bearing carbonate rock. This discovery prompted a drill-coring program, which recovered carbonate rocks from depths of 665 ft (203 m) to 1,000 ft (305 m) (Carlson and Treves, 2005). Core drilling into the geophysical anomaly during the 1970s and 1980s by the State of Nebraska, Cominco American, and Molycorp, Inc. resulted in at least 113 core holes. Molycorp completed 106 of the test holes, recovering about 80,000 ft (24,384 m) of cores and rotary samples (Carlson and Treves, 2005). The deepest hole reached a depth of 3,406 ft (1,038 m) and bottomed in carbonatite.

As part of his doctoral dissertation project, Xu (1996) examined 5,927 ft (1,807 m) of core obtained from the carbonatite complex. He reported that the core studied comprised mostly massive to brecciated, apatite- and pyrochlore-bearing dolomitic carbonatite (89 percent), along with fenitized basalt, lamprophyre, and syenite (totaling 11 percent). Major-element analyses suggest that the carbonate mass is a magnesiocarbonatite, generally similar in major chemical composition to the Iron Hill (Powderhorn) carbonatite stock in southwestern Colorado. The REE are hosted principally by the minerals bastnasite, parisite, and synchisite and by smaller amounts of monazite (Xu, 1996). Niobium resides in pyrochlore. The U.S. Geological Survey obtained a potassium-argon age on biotite in the carbonatite of 544±7 million years old (Xu, 1996).

On May 4, 2010, Quantum Rare Earth Developments Corp. announced that it had acquired the Elk Creek carbonatite properties (*http://www. quantumrareearth. com/)*. In its press release, Quantum reported several assay results from Molycorp's earlier drilling program, which included: detailed drilling of 25 holes within a core (central) zone of the 7-km (4.3- mi) diameter geophysical anomaly and holes spaced about 610 m (2,000 ft) apart surrounding the core zone. Reportedly, the drilling within the core zone found high-grade niobium contents, estimated at "39.4 million tons of 0.82 percent Nb2O5 and is open to the north, west and at depth (Molycorp, Inc. internal memorandum, Feb 05/1 986)." In the widely spaced drilling surrounding the core zone, "least 18 of the surrounding holes intersected greater than 20 feet (6.1 metres) of greater than 1.0% REO [total rare earth oxides], while at least 17 of the surrounding holes intersected greater than 10 feet (3.05 metres) of greater than 0.6% Nb_2O_5." Quantum reports assays intervals that range from 1.02 to 3.12 percent total rare earth oxide. They also note, "Most of the historic drill core, sample rejects, and pulps from Molycorp's exploration are available for review and sampling."

NEW MEXICO—CAPITAN MOUNTAINS

Location: Thin veins containing thorium and rare earth elements crop out on the south flank of the Capitan Mountains in Lincoln County, south-central New Mexico. Latitude: 33.61059 N., Longitude: 105.4505 1 W.; datum: WGS84

Deposit type and basic geology: The radioactive deposits in this district are veins composed of angular fragments of alaskite cemented by quartz. Staatz (1974) identified 12 breccia veins in the district, ranging from 10 to 150 ft (3 to 46 m) in length and 1/4 in. to 8 ft (6 mm to 2.4 m) in thickness. The principal thorium-bearing mineral in these veins is thought to be allanite, accompanied by considerable quantities of quartz, purple fluorite, limonite, and possibly tourmaline (Griswold, 1959).

Status: Currently (2010), there appears to be no active exploration in this district. The deposits were apparently discovered in the early 1 950s during the era of extensive prospecting for radioactivity anomalies. Radioactive veins in the Capitan Mountains were prospected in the middle to late 1 950s by numerous bulldozer cuts but never further developed (Griswold, 1959). The primary focus of the late 1950s exploration of the veins of this district was its thorium potential. A thorium mill was constructed by New Mexico Thorium Company, but it never processed ore (McLemore, 1983). The ruins of the mill were subsequently removed by the U.S. Forest Service.

Production: No mineral resources have been produced from these vein deposits.

Estimated resources: Thorium and REE resources in the district have not been estimated. Staatz (1974) analyzed 17 samples of these veins and found thorium contents of less than 0.01 to as much as 1.12 percent. Reportedly, some assays of vein material showed as much as 1.7 percent thorium (Griswold, 1959). Thorium was assayed as the target commodity in the breccia veins of the southern Capitan Mountains, but REE concentrations are likely to coexist in these deposits. McLemore and others (1988, p. 4) noted that a “select sample assayed 2,500 ppm La, 4,350 ppm Ce, and 330 ppm Y.”

Detailed Discussion

Thin veins containing thorium and rare earth elements (REE) crop out on the south flank of the Capitan Mountains in Lincoln County, south-central New Mexico. The deposits were apparently discovered in the early 1 950s during the era of extensive prospecting for radioactivity anomalies. Radioactive veins in the Capitan Mountains were prospected in the middle to late 1 950s by numerous bulldozer cuts but never further developed (Griswold, 1959). The radioactivity in the veins originates primarily in thorium and in much lesser amounts of uranium.

The radioactive deposits in this district are veins composed of angular fragments of alaskite cemented by quartz. Staatz (1974) identified 12 breccia veins in the district, ranging from 10 to 150 ft (3 to 46 m) in length and 1/4 in. to 8 ft (6 mm to 2.4 m) in thickness. He analyzed 17 samples of these veins and found thorium contents of less than 0.01 to 1.12 percent. Reportedly some assays of vein material showed as much as 1.7 percent thorium (Griswold, 1959). The principal thorium-bearing mineral in these veins is thought to be allanite, which is accompanied by considerable quantities of quartz, purple fluorite, limonite, and possibly tourmaline (Griswold, 1959).

The primary focus of the late 1950s exploration of this district’s veins was their thorium potential. A thorium mill was constructed by New Mexico Thorium Company, but it never processed ore (McLemore, 1983). The ruins of the mill were subsequently removed by the U.S. Forest Service.

Thorium was assayed as the target commodity in the breccia veins of the southern Capitan Mountains, but REE concentrations are likely to coexist in these deposits. McLemore and others (1988, p. 4) noted that a “select sample assayed 2,500 ppm La; 4,350 ppm Ce; and 330 ppm Y.”

New Mexico—El Porvenir District

Location: The El Porvenir or Hermit Mountain district lies about 24 km (15 mi) northwest of Las Vegas and 4.8 km (3 mi) north of Porvenir, on the eastern edge of the Las Vegas Range, San Miguel County, north-central New Mexico. Latitude: 35.74237 N., Longitude: 105.42377 W.; datum: WGS84

Deposit type and basic geology: The bulk of Hermit Mountain is formed by a pink, coarse-grained Precambrian granite that is cut by pegmatite dikes and quartz veins (Robertson, 1976). Some of the pegmatites reportedly contain monazite and rare earth elements mineralization. Little published information is available on the chemistry of these pegmatites, but the data that are available suggest that anomalous rare earth elements concentrations are present.

Status: Currently (2010), there appears to be no active exploration in this district.

Production: No mineral resources have been produced from these occurrences.

Estimated resources: Thorium and rare earth elements resources in the district have not been estimated. McLemore and others (1988) report that samples of "quartzite" contain 546 parts per million (ppm) thorium (Th), 582 ppm lanthanum (La), and 1,160 ppm yttrium (Y). These data presumably refer to a quartz-rich pegmatite.

Detailed Discussion

The El Porvenir or Hermit Mountain district lies about 24 km (15 mi) northwest of Las Vegas and 4.8 km (3 mi) north of Porvenir, on the eastern edge of the Las Vegas Range, San Miguel County, north-central New Mexico. The bulk of Hermit Mountain is formed by a pink, coarse-grained Precambrian granite, which is cut by pegmatite dikes and quartz veins (Robertson, 1976). Some of the pegmatites reportedly contain monazite and rare earth elements mineralization. Little published information is available on the chemistry of these pegmatites, but the data that are available suggest that anomalously high rare earth elements concentrations are present. For example, McLemore and others (1988) report that samples of "quartzite" contain 546 parts per million (ppm) Th, 582 ppm La, and 1,160 ppm Y. These data presumably refer to a quartz- rich pegmatite.

New Mexico—Gallinas Mountains

Location: Gallinas mining district lies in the Gallinas Mountains, about 16 km (10 mi) west of the town of Corona in Lincoln County, central New Mexico. Latitude: 34.19368 N., Longitude: 105.73744 W.; datum: WGS84

Deposit type and basic geology: The cerium-rich mineral bastnasite precipitated in fluorite-copper sulfide deposits in the Gallinas Mountains. The fluorite-copper-bastnasite deposits form veins and fill brecciated zones in sandstones and siltstones of the Permian Yeso Formation. Only two fluorspar deposits were found in porphyritic trachyte; all other fluorspar deposits in the district are hosted by sandstone and siltstones of the Yeso Formation. The porphyritic character of the trachyte and the character of the mineral deposits (low-temperature mineral assemblage, brecciation, infilling of open spaces) suggest that the intrusions and mineralizing events were shallow (hypabyssal). The fluorite-copper-bastnasite deposits of the Gallinas district are found in two settings within the sandstones and siltstones of the Yeso Formation: as veins that fill thin fissures (1–3 in. (2.5–7.6 cm) wide] in highly fractured zones, and as mineral-rich masses that fill open spaces and veinlets in breccia zones. The brecciated zones were formed by faulting and are therefore regarded as fault breccias. In both settings, fluorite is the most abundant mineral.

Status: Currently (2010), there appears to be no active exploration in this district.

Production: In 1953–54, the Gallinas mining district produced small amounts of fluorspar-rich ore (for fluorine); output is estimated to have been less than 2,000 tons (1,800 metric tons) (Griswold, 1959). During 1954–55, the Conqueror No. 9 claim produced approximately 60 tons (54 metric tons) of bastnasite (a cerium-rich mineral) concentrate from these same fluorspar deposits, which was processed nearby in Gallinas at a small mill owned by the United States Rare Earths, Inc. (Griswold, 1959). In 1956, the New Mexico Copper Corp. produced about 300 tons (270 metric tons) of copper-lead-fluorspar ore from its Conqueror claim in the Gallinas district, of which about 11 tons (10 metric tons) of bastnasite ore concentrate from the Conqueror No. 10 claim.

Estimated resources: Soule (1946) estimated that bastnasite forms about 5 percent of the breccia deposits. Soule (1946) analyzed hand-picked grains of bastnasite from the Gallinas district and found that the bastnasite contained 74.39 percent total rare earth elements oxides: 25.61 percent cerium oxide and 48.78 percent other rare earth elements oxides. No resource estimate of the potential fluorspar and bastnasite tonnage within the Gallinas district has been published.

Detailed Discussion

The cerium-rich mineral bastnasite formed within fluorite-copper sulfide deposits in the Gallinas Mountains, about 16 km (10 mi) west of the town of Corona in Lincoln County, central New Mexico. In 1953–54, the Gallinas mining district produced small amounts of fluorspar-rich ore (for fluorine); output is estimated to have been less than 2,000 tons (1,800 metric tons) (Griswold, 1959). During 1954–55, the Conqueror No. 9 claim produced approximately 60 tons (54 metric tons) of bastnasite concentrate from these same fluorspar deposits, which was processed nearby in Gallinas at a small mill owned by the United States Rare Earths, Inc. (Griswold, 1959). In 1956, the New Mexico Copper Corp. produced about 300 tons (270 metric tons) of copper-lead-fluorspar ore from its Conqueror claim in the Gallinas district, of which about 11 tons (10 metric tons) was bastnasite ore concentrate from the Conqueror No. 10 claim.

The Gallinas Mountains consist of Lower Permian sedimentary rocks that were domed, uplifted, faulted, and fractured during the emplacement of middle(?) Tertiary laccoliths composed mainly of alkaline trachyte and rhyolite (Perhac, 1970). The fluorite-copper-bastnasite deposits form veins and fill brecciated zones in sandstones and siltstones of the Permian Yeso Formation. Only two fluorspar deposits were found in porphyritic trachyte; all other fluorspar deposits in the district are hosted by sandstone and siltstones of the Yeso Formation. The district's fluorite-copper-bastnasite deposits are interpreted to result from epithermal (relatively low temperature) mineralization related to the intrusion of the alkaline trachyte (Perhac and Heinrich, 1964; Perhac, 1970). The porphyritic character of the trachyte and the character of the mineral deposits (low-temperature mineral assemblage, brecciation, and infilling of open spaces) suggest that the intrusions and mineralizing events occurred at shallow depths (hypabyssal).

The fluorite-copper-bastnasite ore of the Gallinas district was deposited in two settings within the sandstones and siltstones of the Yeso Formation: as veins that fill thin fissures (1–3 inches (2.5–7.6 cm wide) in highly fractured zones, and as mineral-rich masses that fill open spaces and veinlets in breccia zones. The brecciated zones were formed by faulting and are therefore regarded as fault breccias. In both settings, fluorite is the most abundant mineral. Fluorite content in the breccia deposits averages about 60 percent (Soule, 1946). Barite, calcite, and quartz are next in abundance (Soule, 1946; Perhac, 1970). Other associated minerals are pyrite, galena, chalcopyrite, mimetite, sphalerite, conichalcite, chalcocite, wulfenite, malachite, azurite, vanadinite, mottramite, cerussite, chrysocolla, agardite (yttrium-bearing mineral), and bastnasite (Glass and Smalley, 1945; Griswold, 1959; Perhac, 1970; DeMark, 1980).

The bastnasite ore forms thin, tabular, waxy yellow, transparent to translucent crystals 1 to 10 mm in length, usually about 4 mm (0.16 in.) in width and embedded in fluorite (Glass and Smalley, 1945; Soule, 1946; DeMark, 1980). Soule (1946) estimated that bastnasite forms about 5 percent of the breccia deposits. Soule (1946) analyzed hand-picked grains of bastnasite from the Gallinas district and found that the bastnasite contained 74.39 percent total rare earth elements oxides: 25.61 percent cerium oxide and 48.78 percent other rare earth elements oxides. No resource estimate of the potential fluorspar and bastnasite tonnage within the Gallinas district has been published.

New Mexico—Gold Hill Area and White Signal District

Location: The White Signal district is located in Grant County, southwestern New Mexico. The adjacent Gold Hill area lies near the crest of the Burro Mountains. Latitude: 32.45400 N., Longitude: 108.50603 W.; datum: WGS84

Deposit type and basic geology: Rare earth elements (REE)– thorium–bearing minerals form pods and lenses within pegmatites hosted by the Proterozoic Burro Mountain granite in the western part of the White Signal district. Quartz, muscovite, and microcline are the primary minerals of these pegmatites. Large euhedral crystals of euxenite [$Y,Ca,Ce,U,Th)(Nb,Ta,Ti)_2O_6$] are found locally, and some crystals are several inches long (Gillerman, 1964). Other REE-bearing minerals reported in the pegmatites are allanite and samarskite (Richter and others, 1986). In the Gold Hill area, near the crest of the Burro Mountains, the same REE-bearing minerals are hosted in similar but larger pegmatites that also cut the Burro Mountain granite (Hedlund, 1978). The primary pegmatite minerals are milky quartz, microcline, albite, and muscovite with accessory biotite, magnetite, garnet, fluorite, and REE-bearing minerals such as allanite, euxenite, and samarskite.

Status: Currently (2010), there appears to be no active exploration in this district.

Production: Shallow prospect pits were dug into the pegmatites in order to explore their radioactivity, presumably during the 1950s, but no further development is reported.

Estimated resources: No report of the REE concentrations in these pegmatites has been published, but thorium concentrations can reach as high as 0.72 percent (Staatz, 1974).

Detailed Discussion

Rare earth elements (REE)–thorium–bearing minerals form pods and lenses within pegmatites hosted by the Proterozoic Burro Mountain granite in the western part of the White Signal district, which lies in Grant County, southwestern New Mexico (Gillerman, 1964; Richter and Lawrence, 1983; Richter and others, 1986). Shallow prospect pits were dug into the pegmatites in order to explore their radioactivity, presumably during the 1950s, but no further development is reported.

Quartz, muscovite, and microcline are the primary minerals of these pegmatites. Large euhedral crystals of euxenite [Y,Ca,Ce,U,Th)(Nb,Ta,Ti)$_2$O$_6$] are found locally, and some crystals are several inches long (Gillerman, 1964). Other REEbearing minerals reported in the pegmatites are allanite and samarskite (Richter and others, 1986). No analyses of the REE concentrations in these pegmatites have been reported.

In the Gold Hill area, near the crest of the Burro Mountains, the same REE-bearing minerals are hosted in similar but larger pegmatites that also cut the Burro Mountain granite (Hedlund, 1978). These pegmatites were prospected by small pits between 1952 and 1955. However, "the amount and concentration of rare-earth minerals was so small that work was soon stopped" (Gillerman, 1964, p. 127). The primary pegmatite minerals are milky quartz, microcline, albite, and muscovite with accessory biotite, magnetite, garnet, fluorite, and the REE-bearing minerals, such as allanite, euxenite, and samarskite. No report of the REE concentrations in these pegmatites has been published, but Th concentrations can reach as high as 0.72 percent (Staatz, 1974).

NEW MEXICO—LAUGHLIN PEAK AREA

Location: Thorium (Th)- and rare earth elements–rich veins are intruded throughout an area of about 12 km^2 (4.6 mi^2) near Laughlin Peak, about 38 km (24 mi) southeast of Raton in Colfax County, northeastern New Mexico. Latitude: 36.58 156 N., Longitude: 104.22953 W.; datum: WGS84

Deposit type and basic geology: In this area, Staatz (1985) mapped 29 veins ranging from 0.5 to 550 m (1.6 to 1,800 ft) in length and 0.2 to 70 cm (0.08 to 28 in.) in thickness. Thorium- and REE-bearing minerals in the veins include brockite, xenotime, and crandallite. The brockite and xenotime are mainly enriched in the yttrium-group (heavy) rare earths, whereas the crandallite contains mostly cerium-group (light) rare earth elements. The veins are steeply dipping and lie along fracture zones, cutting mostly trachyte and Dakota Sandstone but also intrusive breccia and trachyandesite. The gangue minerals are mostly potassium feldspar, quartz, or calcite, and lesser amounts of goethite, magnetite, barite, zircon, rutile, and a manganese oxide. One small carbonatite dike was found 2.7 km (1.7 mi) south of the mapped area.

Status: Currently (2010), there appears to be no active exploration in this district.

Production: Prospecting for radioactive deposits began in the Laughlin Peak area in the early 1 950s; small pits and trenches were dug along the veins. These prospects were subsequently sampled by Staatz (1985).

Estimated resources: Thorium and REE resources in the district have not been estimated. Staatz (1985) found that most of the veins contain higher concentrations of the yttrium-group REE than the cerium group, and veins with high yttrium-group concentrations usually also have a high thorium content. Sampling by Staatz (1985, p. 1) found, "Thorium content of 30 samples ranges from 30 to 24,200 ppm (parts per million), and the total rare-earth content from 147 to 19,030 ppm." These amounts equate to REE concentrations of about 0.018 to 2.34 percent total REE oxide.

Detailed Discussion

Thorium (Th)- and rare earth elements (REE)-rich veins crop out throughout an area of about 12 km^2 (4.6 mi^2) near Laughlin Peak, about 38 km (24 mi) southeast of Raton in Colfax County of northeastern New Mexico. Staatz (1985) mapped 29 veins in this area, ranging from 0.5 to 550 m (1.6 to 1,800 ft) in length and 0.2 to 70 cm (0.08 to 28 in.) in thickness. Thorium- and REE-bearing minerals in the veins include brockite, xenotime, and crandallite. Thorite and monazite are absent. The brockite and xenotime are mainly enriched in the yttrium-group (heavy) rare earths, whereas the crandallite contains mostly cerium-group (light) rare earth elements. Staatz (1985) found that most of the veins contain higher concentrations of yttrium-group REE than cerium-group REE, and veins with high yttrium-group values usually have a high Th content. Sampling by Staatz (1985, p. 1) found, "Thorium content of 30 samples ranges from 30 to 24,200 ppm (parts per million), and the total rare-earth content from 147 to 19,030 ppm." These amounts equate to REE concentrations of about 0.018 to 2.34 percent total REE oxide.

The veins are steeply dipping and lie along fracture zones, cutting mostly trachyte and Dakota Sandstone but also intrusive breccia and trachyandesite. The gangue minerals are mostly potassium feldspar, quartz, or calcite, and lesser amounts of goethite, magnetite, barite, zircon, rutile, and a manganese oxide. One small carbonatite dike was found 2.7 km (1.7 mi) south of the mapped area (Staatz, 1985). The igneous rocks that are spatially associated with the veins have alkaline compositions of phonolite, trachyte, trachyandesite, and basalt. These rocks are also anomalous in REE, especially light REE, and show total REE contents of 173–807 ppm. The veins do not cut phonolite or basalt, but Staatz (1985) suggested that the source of the Th and REE in the veins was the magma that formed the phonolite during the Oligocene, because samples of the phonolite showed Th and REE contents much higher than samples of other associated igneous rocks.

Prospecting for radioactivity began in the Laughlin Peak area in the early 1950s; small pits and trenches were dug along the veins. These prospects were subsequently sampled by Staatz (1985). Additional exploration is necessary to evaluate the full REE resource potential of this area, but the geologic mapping and descriptions by Staatz (1985, 1986, 1987) provide a solid framework for further work here.

NEW MEXICO—LEMITAR AND CHUPADERA MOUNTAINS

Location: Carbonatite dikes and veins occur in the Lemitar Mountains in west-central New Mexico, and more than a dozen similar carbonatite dikes are known to be located to the south within the adjacent Chupadera Mountains. These mountain ranges lie west of San Antonio, Socorro, and Lemitar in Socorro County, New Mexico. Latitude: 34.15398 N., Longitude: 106.98623 W.; datum: WGS84, and Latitude: 33.85285 N., Longitude: 106.9578 1 W.; datum: WGS84

Deposit type and basic geology: More than 100 carbonatite dikes and veins that contain rare earth elements (REE) cut Precambrian metamorphic and granitic terrane in the Lemitar Mountains, and more than a dozen similar carbonatite dikes intruded Precambrian metamorphic rocks to the south within the adjacent Chupadera Mountains (McLemore, 1983, 1987; Van Allen and others, 1986). The carbonatite intrusions range from less than 1 cm (0.4 in.) thick—veins—to more than 1 m (3.3 ft) thick—dikes (McLemore, 1983, 1987). A few of the dikes can be traced in outcrop for as much as 600 m (1,970 ft). Subparallel sets of carbonatites locally form dike swarms. Alkaline igneous rocks are lacking in these mountain ranges, so their igneous source presumably lies at depth (McLemore, 1987). Age determinations by the potassium-argon method suggest that the carbonatites are Ordovician (449±16 million years old, McLemore, 1987) and thus represent a part of widespread Cambrian-Ordovician igneous activity in New Mexico.

Status: Currently (2010), there appears to be no active exploration in this district.

Production: No mineral resources have been produced from these vein deposits. Because uranium and thorium in the carbonatite dikes make the dikes radioactive, they were identified in 1954 during a uranium exploration program conducted by United Geophysical Corp. (Van Allen and others, 1986).

Estimated resources: Thorium and REE resources in the district have not been estimated. McLemore and others (1988) reported a maximum concentration from selected samples of 1,950 parts per million (ppm) thorium (0.195 percent) and 0.25 weight percent uranium oxide. Van Allen and others (1986) and McLemore and others (1988) report maximum concentrations from select carbonatite samples as 0.19 weight percent total REE; 700 ppm Y; 4,900 ppm cerium (Ce); and 1,700 ppm lanthanum (La).

Detailed Discussion

More than 100 carbonatite dikes and veins that contain rare earth elements (REE) cut Precambrian metamorphic and granitic terrane in the Lemitar Mountains in west-central New Mexico, and more than a dozen similar carbonatite dikes intruded Precambrian metamorphic rocks to the south within the adjacent Chupadera Mountains (McLemore, 1983, 1987; Van Allen and others, 1986). These north-south-trending mountain ranges lie west of San Antonio, Socorro, and Lemitar in Socorro County, New Mexico. The carbonatite intrusions range from less than 1 cm (0.4 in.) thick—veins—to more than 1 m (3.3 ft) thick—dikes (McLemore, 1983, 1987). A few of the dikes can be traced in outcrop for as much as 600 m (1,970 ft). Subparallel sets of carbonatites locally form dike swarms.

The carbonatite dikes occupy sets of fractures apparently related to rifting of the adjacent Rio Grande Rift. Alkaline igneous rocks are lacking in these mountain ranges, so the dikes' igneous source presumably lies at depth (McLemore, 1987). Age determinations by the potassium-argon method suggest that the carbonatites are Ordovician (449±16 million years old, McLemore, 1987) and thus represent a part of widespread Cambrian-Ordovician igneous activity in New Mexico.

Because uranium and thorium in the carbonatite dikes make the dikes radioactive, they were identified in 1954 during a uranium exploration program conducted by United Geophysical Corp. (Van Allen and others, 1986). Although they were described at that time as "radioactive-calcite veins" they were later classified as carbonatites in 1978 by Tenneco geologists. McLemore and others (1988) reported a maximum concentration from selected samples of 1,950 ppm thorium (0.195 percent) and 0.25 weight percent U_3O_8. Although the uranium correlates with yttrium in carbonatites of the Chupadera Mountains, no specific uranium or yttrium mineral was identified (McLemore, 1983; Van Allen and others, 1986). The gangue mineralogy of these carbonatites is detailed by McLemore (1983, 1987).

Overall, the REE concentrations found in the carbonatite dikes in both mountain ranges were quite variable. Van Allen and others (1986) and McLemore and others (1988) report maximum concentrations from select carbonatite samples as 0.19 weight percent total REE; 700 ppm Y; 4,900 ppm cerium (Ce); and 1,700 ppm lanthanum (La).

New Mexico—Petaca District

Location: The Petaca district is located between Ojo Caliente and Tres Piedras, in Rio Arriba County, north-central New Mexico. Latitude: 36.58835 N., Longitude: 106.07170 W.; datum: WGS84

Deposit type and basic geology: Thorium- and rare earth elements (REE)–bearing pegmatites crop out in Precambrian rocks in the southeastern Tusas Mountains (Bingler, 1968). The pegmatites of the Petaca district take a variety of shapes, such as dikes, sills, pipes, pods, troughs, and irregular forms. The pegmatites crop out for 75 to 1,430 ft (23 to 436 m) in length (an average outcrop length is 410 ft (125 m)), and they have an average width of 30 to 35 ft (9 to 11 m) (Bingler, 1968). Elevated REE concentrations in Petaca district pegmatites mainly reflect the mineral samarskite, a REE-iron-uranium-thorium-niobium-tantalum-titanium–bearing oxide.

Status: Currently (2010), there appears to be no active exploration in this district.

Production: A number of pegmatites in the district were mined for their large books of muscovite mica crystals, beginning in 1870 and continuing intermittently until 1944 (Bingler, 1968).

Estimated resources: Thorium and REE resources in the district have not been estimated. McLemore and others (1988) reported an average niobium content of 0.04 percent in 87 pegmatites of the district. Monazite accounts for the pegmatites' high thorium content. An analysis of the Globe pegmatite in the district found 10,332 ppm thorium (McLemore and others, 1988). McLemore and others (1988, p. 4) reported the following REE analysis of a sample of the Globe pegmatite: "600 ppm Y, 660 ppm Yb, 396 ppm Er, 186 ppm Gd, 3,117 ppm [total] REE + Y." Otherwise, the REE content of the pegmatites of the Petaca district has not been published.

Detailed Discussion

Thorium- and rare earth elements (REE)–bearing pegmatites are exposed in the Petaca district, located between Ojo Caliente and Tres Piedras, in Rio Arriba County, north- central New Mexico. The pegmatites crop out in Precambrian rocks in the southeastern Tusas Mountains (Bingler, 1968). The pegmatites of the Petaca district take a variety of shapes, such as dikes, sills, pipes, pods, troughs, and irregular forms. The pegmatite forms and their characteristics are described in detail by Jahns (1946). They crop out for 75 to 1,430 ft (23 to 436 m) in length (an average outcrop length is 410 ft (125 m)), and they have an average width of 30 to 35 ft (9 to 11 m) (Bingler, 1968).

The primary minerals of the Petaca district pegmatites are microcline, quartz, plagioclase, and muscovite. A number of pegmatites in the district were mined for their large books of muscovite mica crystals beginning in 1870 and continuing intermittently until 1944 (Bingler, 1968). Almost 50 accessory minerals have been identified; the most common is spessartite (now called spessartine) garnet, columbite-tantalite [$(Fe,Mn)(Nb,Ta)_2O_6$], fluorite, beryl, monazite, samarskite, and ilmenite-magnetite (Wright, 1948; Redmon, 1961; Bingler, 1968). (Columbite and tantalite are obsolete names for a mineral series; columbite is now named ferrocolumbite ($Fe^{2+}Nb_2O_6$), which forms two minerals series, with ferrotantalite ($Fe^{2+}Ta_2O_6$) and with manganocolumbite [(Mn^{2+},Fe^{2+}) $(Nb,Ta)_2O_6$)].

The elevated concentrations of niobium (Nb) and tantalum (Ta) in the pegmatites arise from Nb- and Ta-bearing minerals. For example, McLemore and others (1988) reported an average of Nb content of 0.04 percent in 87 pegmatites of the district. Similarly, monazite accounts for the pegmatites' high Th content. Specifically, analysis of the Globe pegmatite in the district found 10,332 ppm Th (McLemore and others, 1988).

Elevated REE concentrations in Petaca district pegmatites mainly reflect the mineral samarskite, an REE-iron-uraniumthorium-niobium-tantalum-titanium–bearing oxide. The REE are reportedly restricted to albite-rich zones in the pegmatites. McLemore and others (1988, p. 4) reported this REE analysis of a sample of the Globe pegmatite: "600 ppm Y, 660 ppm Yb, 396 ppm Er, 186 ppm Gd, 3,117 ppm [total] REE + Y." Otherwise, the REE content of the pegmatites of the Petaca district has not been published.

New Mexico—Red Hills Area

Location: Dike-like and tabular bodies containing thorium and rare earth elements are exposed in the Red Hills area of the southern Caballo Mountains, Sierra County, New Mexico. They crop out across an area of about 7.8 square km (3 mi^2), which is centered about 4 km (2.5 mi) southeast of Caballo dam. Latitude: 32.86293 N., Longitude: 107.25655 W.; datum: WGS84

Deposit type and basic geology: At least 45 radioactive, dike-like and tabular, deep-red bodies of syenite that crop out in the Red Hills area are modestly enriched in heavy rare earth elements (HREE). These coarse-grained, microcline-rich syenites range from 1 to 100 m (3.3 to 328 ft) in length and several centimeters to 10 m (1 in. to 33 ft) in width (McLemore, 1986). The syenite bodies are composed mainly of microcline and contain lesser amounts of quartz, muscovite, hematite, goethite, chlorite, and plagioclase and accessory apatite, zircon, calcite, fluorite, limonite, magnetite, and barite. The radioactivity originates in uranium and thorium residing in with spinel, rutile, anatase, thorite, thorogummite, and possibly uraninite (Staatz and others, 1965; McLemore, 1986).

Status: Currently (2010), there appears to be no active exploration in this district.

Production: No mineral resources have been produced from these vein deposits.

Estimated resources: Thorium or REE resources in the district have not been estimated. Subsurface sampling is necessary in this area to determine if a larger syenite mass exists at depth, and if such a mass is consistently enriched in the heavy REE. Samples of the microclinerich (syenite) bodies in the Red Hills by Staatz and others (1965) contained thorium concentrations as much as 0.44 weight percent and modest to undetectable concentrations of the light REE. However, the proportion of heavy REE was higher. In particular, yttrium concentrations were as much as 0.19 weight percent (Staatz and others, 1965; McLemore, 1986).

Detailed Discussion

At least 45 radioactive, dike-like and tabular, deep-red bodies of syenite that are modestly enriched in heavy rare earth elements (REE) are exposed in the Red Hills area of the southern Caballo Mountains, Sierra County, New Mexico (Staatz and others, 1965; McLemore, 1983, 1986). These coarse-grained, microcline-rich syenites range from 1 to 100 m (3.3 to 328 ft) in length and several centimeters to 10 m (1 in. to 33 ft) in width (McLemore, 1986). They crop out across an area of about 7.8 square km (3 mi^2), which is centered about 4 km (2.5 mi) southeast of Caballo dam. The syenite bodies are composed mainly of microcline, with lesser amounts of quartz, muscovite, hematite, goethite, chlorite, and plagioclase, and accessory apatite, zircon, calcite, fluorite, limonite, magnetite, and barite. The radioactivity originates in uranium and thorium that resides in with spinel, rutile, anatase, thorite, thorogummite, and possibly uraninite (Staatz and others, 1965; McLemore, 1986).

Samples of the microcline-rich (syenite) bodies in the Red Hills by Staatz and others (1965) contained thorium concentrations as much as 0.44 weight percent and modest to undetectable concentrations of the light REE. However, the proportion of heavy REE was higher. In particular, yttrium concentrations were as much as 0.19 weight percent (Staatz and others, 1965; McLemore, 1986). Subsurface sampling is necessary in this area to determine if a larger syenite mass exists at depth, and if such a mass is consistently enriched in the heavy REE.

New Mexico—Wind Mountain, Cornudas Mountains

Location: Wind Mountain is located in Otero County, New Mexico, and is one of the largest uplifted areas of the Cornudas Mountains. Wind Mountain stands about 80 km (50 mi) east of El Paso, just north of the New Mexico–Texas boundary. Latitude: 32.02382 N., Longitude: 105.50162 W.; datum: WGS84

Deposit type and basic geology: Wind Mountain was formed by a laccolith of porphyritic nepheline syenite that rises about 2,500 ft (762 m) above the surrounding Diablo Plateau (Holser, 1959). Dikes and sills of nepheline syenite and syenite cut the main mass of the laccolith. At least some of these dikes and sills contain thorium, uranium, and rare earth elements (REE) mineralization (McLemore, 1983). The alkaline dikes and sills reportedly also contain anomalous concentrations of beryllium (Be), niobium (Nb), lithium (Li), nickel (Ni), tin (Sn), zirconium (Zr), and fluorine (F, in fluorite).

Status: Currently (2010), there appears to be no active exploration in this district.

Production: No mineral resources have been produced from these dikes and sills.

Estimated resources: Thorium and REE resources in the district have not been estimated. McLemore and others (1988) analyzed a dike sample collected from Wind Mountain and reported 700 parts per million (ppm) lanthanum (La), 270 ppm neodymium (Nd), and 242 ppm yttrium (Y). However, a full rare earth elements resource evaluation of the Wind Mountain uplift would require much more sampling than has been conducted thus far.

Detailed Discussion

Wind Mountain, in Otero County, New Mexico, is one of the largest uplifted areas of the Cornudas Mountains, a mountain range that straddles the New Mexico–Texas border east of El Paso, Texas. The Cornudas Mountains, the northern end of an alkaline magmatic belt that was emplaced about 35 million years ago, extends from southern New Mexico, across Texas, and into Mexico. Wind Mountain itself lies about 80 km (50 mi) east of El Paso, just north of the New Mexico- Texas boundary.

Wind Mountain was formed by a large alkaline intrusion, a laccolith of porphyritic nepheline syenite that rises about 2,500 ft (762 m) above the surrounding Diablo Plateau (Holser, 1959). Dikes and sills of nepheline syenite and syenite cut the main mass of the laccolith. At least some of these dikes and sills contain Th, U, and rare earth elements mineralization (McLemore, 1983). The alkaline dikes and sills reportedly also contain anomalously high concentrations of beryllium (Be), niobium (Nb), lithium (Li), nickel (Ni), tin (Sn), zirconium (Zr), and fluorine (F, in fluorite). McLemore and others (1988) analyzed a dike sample collected from Wind

Mountain and reported concentrations of 700 ppm lanthanum (La), 270 ppm neodymium (Nd), and 242 ppm yttrium (Y). However, a full rare earth elements resource evaluation of the Wind Mountain uplift would require much more sampling than has been completed thus far.

NEW YORK—MINEVILLE IRON DISTRICT

Location: The Mineville iron district includes the iron ores once mined in the Mineville, New York, area, located in the northeastern part of the Adirondack Mountains, on the west side of Lake Champlain. Most of the former iron mines are near the towns of Mineville and Port Henry in Essex County, New York. This district of iron deposits extends for approximately 78 km^2 (30 mi^2). Latitude: 44.06403 N., Longitude: 73.49239 W.; datum: WGS84

Deposit type and basic geology: Thorium and rare earth elements (REE) reside within apatite in iron ores once mined in the Mineville, New York, area. The primary apatite-rich iron deposits are the Old Bed, Cheever, and Smith bodies; the Cheever and Smith orebodies have been mined out. The orebodies are magnetite deposits that are intricately folded and faulted within a complex suite of Precambrian metamorphic and igneous rocks. The host rocks have both mafic and felsic compositions that include augite syenites, granite, gabbro and diorite (Kemp, 1908; Staatz and others, 1980). The iron deposits are mainly magnetite, martite, and apatite, with gangue minerals of augite, hornblende, albite, quartz, pyrite, and tourmaline (McKeown and Klemic, 1956). The iron deposits in the Mineville–Port Henry area that are high in apatite are also enriched in phosphorous, thorium, and REE, because these elements are concentrated within the apatite grains. In addition, the ore mineral magnetite is intergrown with 1–3 mm (0.04–0.12 in.) long, rice-shaped grains of apatite.

Status: Currently (2010), there is no reported exploration or development in this district.

Production: Iron ore was mined from the district intermittently from 1804 until the last operation closed in 1971. A detailed mining history of the district is summarized by Staatz and others (1980).

Estimated resources: Currently, large tailings piles and unmined parts of magnetite orebodies in the Mineville district contain REE-bearing apatite-rich rock. Staatz and others (1980) estimated that about two-thirds of the tailings piles were derived from apatite-rich ores, which would represent about 9 million metric tons (10 million tons) of the tailings. Using an average grade of about 8 percent apatite content, approximately 720,000 metric tons (790,000 tons) of apatite could be present in the tailings dumps in the district. McKeown and Klemic (1956) reported an average rare earth oxide content of 11.14 percent in 14 samples of apatite separated from the Old Bed, Joker, and Smith orebodies. Thus, the tailings dump piles could contain approximately 80,200 metric tons (88,400 tons) of rare earth oxides.

Detailed Discussion

Thorium and rare earth elements (REE) are incorporated within apatite in iron ores once mined in the Mineville, New York, area, located in the northeastern part of the Adirondack Mountains, on the west side of Lake Champlain (figure 16). Most of the former iron mines are near the towns of Mineville and Port Henry in Essex County, New York. This district of iron deposits extends for approximately 78 km^2 (30 mi^2). The primary apatite-rich iron deposits are the Old Bed, Cheever, and Smith bodies; the Cheever and Smith orebodies have been mined out. Iron ore was mined from the district intermittently from 1804 until the last operation closed in 1971. A more detailed mining history of the district is summarized by Staatz and others (1980).

The orebodies are magnetite deposits that are intricately folded and faulted within a complex suite of Precambrian metamorphic and igneous rocks. The host rocks have both mafic and felsic compositions that include augite syenites, granite, gabbro, and diorite (Kemp, 1908; Staatz and others, 1980). The granite has felsic Na- and K-rich compositions that alternate with more basic, amphibole-, pyroxene-, and phlogopite-bearing rocks. Overlying the igneous sequence rests a metasedimentary series that contains Proterozoic marbles, calc-silicates and gneisses. The iron deposits are mainly magnetite, martite, and apatite, with gangue minerals of augite, hornblende, albite, quartz, pyrite, and tourmaline (McKeown and Klemic, 1956). Pegmatites crosscut the magnetite ore and consist of quartz, feldspar ± magnetite ± allanite, and minor scapolite, titanite, epidote, and zircon.

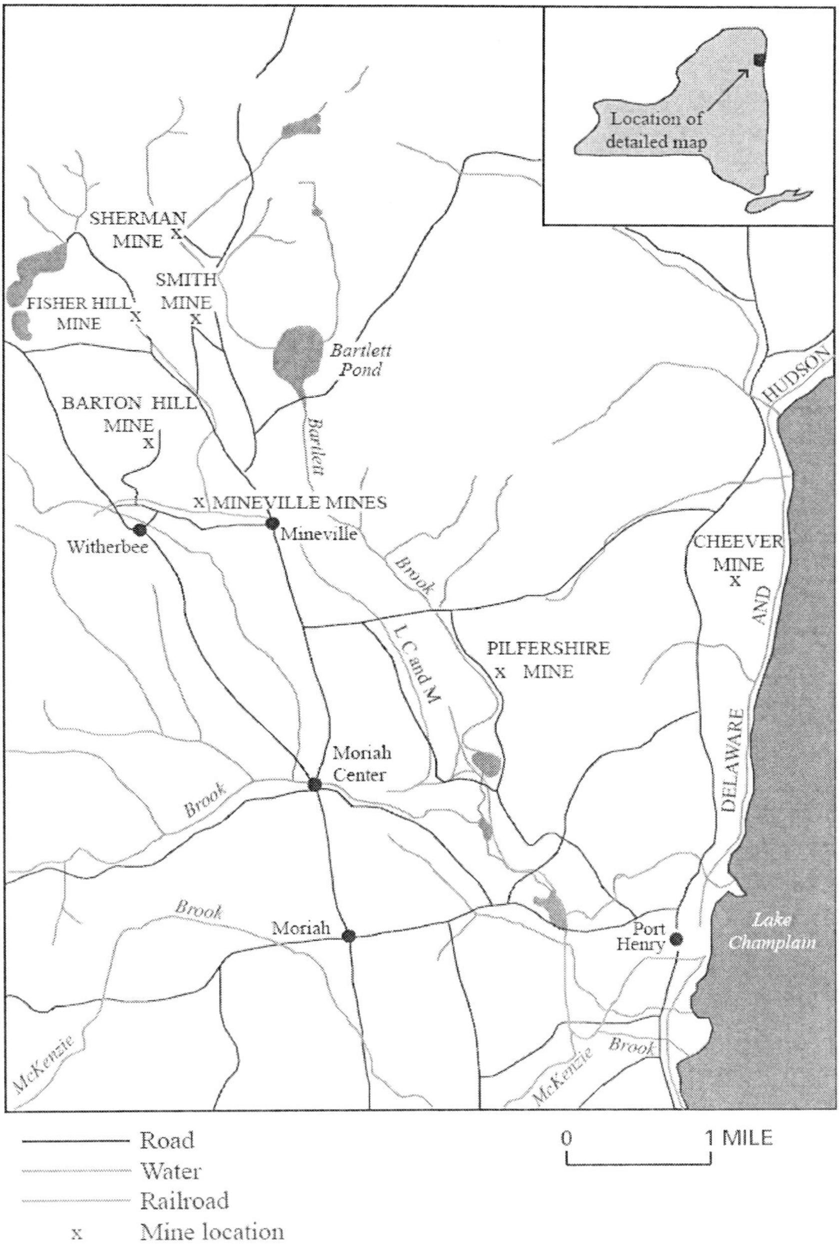

Figure 16. Mineville district in Essex County, New York. Modified from McKeown and Klemic (1956).

The iron deposits in the Mineville–Port Henry area that are high in apatite content are also enriched in phosphorous, thorium, and REE, because these elements are concentrated within the apatite grains. In addition, the ore mineral magnetite is intergrown with 1–3 mm (0.04–0.12 in.) long, rice-shaped grains of apatite. The apatites can take several colors, such as reddish brown, green, white, or transparent. The reddish-brown variety is the most common. The reddish- brown color of the apatite, also referred to as fluorapatite, is most likely generated by infiltration or inclusions of hematite along fractures or within the crystal structure.

According to Staatz and others (1980, p. 29), "The reddish-brown apatite contains between 5.8 and 20.6 percent total rare earths, the green variety between 0.5 and 2.0 percent, and the white and transparent varieties only trace amounts." Monazite, bastnasite, and hematite fill microfractures in the apatites and also form coatings on the apatites as well. In addition, microscopic phases of secondary thorite, allanite, and parisite have been noted in some apatite crystals. Monazite, thorite, allanite, and bastnasite are enriched in thorium and REE. Kainosite has also been observed in edenite under a polarizing microscope and scanning electron microscope (Lupulescu and Pyle, 2008).

Allanite located in the pegmatite bodies, host gneiss, or pyroxene-rich rocks is rich in cerium. The allanite crystals with a pegmatitic origin are very large, 20 to 25 cm (7.9 to 9.8 in.) long, 6 to 20 cm (2.4 to 7.9 in.) wide, and 2.5 to 5 cm (1 to 2 in.) thick, with smooth surfaces and conchoidal fractures (Blake, 1858). Allanite crystals associated with quartz and monazite (Ce) are large, smooth-faced, and metamict (Lupulescu and Pyle, 2008). In the pyroxene rocks, the metamict allanite has decomposed to monazite (Ce), which contains a rim of Y-dominant allanite that is dark brown and strongly pleochroic.

In the iron ores of the Mineville–Port Henry area, two generations of monazite can be distinguished largely on the basis of their relationships with other minerals. Within the pegmatites, monazite (Ce) ranges from 1–3 mm (0.04–0. 12 in.) to almost 1 cm (0.39 in.) and appears to be associated with or inclusions in allanite, both of which are embedded in quartz (Lupulescu and Pyle, 2008). Monazite also formed as a minute, secondary mineral because of the breakdown of allanite in the pyroxene-rich rocks.

Currently, large tailings piles and unmined parts of magnetite orebodies in the Mineville district contain REE-bearing apatite-rich rock. Staatz and others (1980) estimated that about two-thirds of the tailings piles were derived from apatite-rich ores, which would represent about 9 million metric tons of the tailings. Using an average grade of about 8 percent apatite, approximately

720,000 metric tons of apatite could be present in the tailings dumps in the district. McKeown and Klemic (1956) reported an average rare earth elements–oxide content of 11. 14 percent in 14 samples of apatite separated from the Old Bed, Joker, and Smith orebodies. Thus, the tailings dump piles could contain approximately 80,200 metric tons of rare earth oxides. Uranium and thorium contents average 0.032 percent and 0. 15 percent, respectively. Spectrographic analyses also reveal that yttrium is one of the main REE contained in apatite, making the tailings a potential source for the yttrium-group REE (McKeown and Klemic, 1956).

Wyoming—Bear Lodge Mountains

Location: Rare earth elements (REE)–thorium deposits are exposed in the southern Bear Lodge Mountains, about 8 km (5 mi) northwest of Sundance, Crook County, Wyoming. Latitude: 44.49215 N., Longitude: 104.44133 W.; datum: WGS84

Deposit type and basic geology: The REE-thorium deposits and nearby gold mineralization of the southern Bear Lodge Mountains are hosted by middle Tertiary alkaline intrusions. These intrusions are Eocene (38.3–50 million years old) and consist primarily of phonolite and trachyte (Staatz, 1983). They intruded Paleozoic and Mesozoic sedimentary rocks, forming a dome about 13 km (8 mi) long by 10 km (6 mi) wide. Breccia bodies are associated with the igneous intrusions, such as a heterolithic diatreme breccia near Bull Hill. Rare earth elements–bearing carbonatite dikes intruded near the Bull Hill diatreme; the dikes are surrounded by a large zone of low-grade REE mineralization that fills thin, narrow stockwork fractures within the large alkaline intrusions. These thorium and REE deposits crop out throughout an area of about 16 km^2 (6 mi^2) (Staatz, 1983). The igneous core of the dome is microfractured and altered, thereby forming dis seminated deposits. The REE and thorium mineralization precipitated within thin fractures as coatings and veinlets as much as 6 mm thick. The coatings and veinlets consist predominantly of iron and manganese oxide minerals, along with potassium feldspar and quartz. The REE and thorium are incorporated into the minerals monazite, thorite, and brockite.

Status: During the last few field seasons, Rare Element Resources, Ltd., has explored for REE (http://www. rareelementresources.com/s/Home.asp), focused primarily on Bull Mountain (near the center of the dome) and areas just to the west and southeast of Bull Mountain, where numerous carbonatite dikes are exposed. The company drilled again in the summer months of 2010.

Production: No mineral resources have been produced from these vein deposits thus far (2010).

Estimated resources: Rare Element Resources, Ltd., has delineated three mineralogical zones in the Bear Lodge Mountain deposits dependent on depth of weathering— oxide, transitional, and unoxidized. As of 2010, they estimate that the oxide zone—the near-surface part of the "Bull Hill deposit"—hosts 4.5 million tons at 4.3 percent REE oxides. They suggest that the total inferred resource of the deposit is 9.8 million tons averaging 4.1 percent REE oxides. During 2009, the company drilled five holes just northwest of Bull Mountain through a total of 5,141 vertical feet of veins and dikes; there, total REE oxide concentrations ranged from 2.08 to 9.12 percent.

Detailed Discussion

Disseminated and vein rare earth elements (REE)– thorium deposits are located in the southern Bear Lodge Mountains, about 8 km (5 mi) northwest of Sundance, Crook County, Wyoming. These REE deposits have been the focus of recent exploration and resource evaluation by Rare Element Resources, Ltd. (http://www. rareelementresources. com/s/Home.asp). Three decades ago, the geology and thorium-REE resources of this district were studied by the USGS (Staatz, 1983).

The REE-thorium deposits and nearby gold (Au) mineralization of the southern Bear Lodge Mountains are hosted by middle Tertiary alkaline intrusions. These intrusions are Eocene (38.3–50 million years old) and consist primarily of phonolite and trachyte (Staatz, 1983). They intruded Paleozoic and Mesozoic sedimentary rocks, forming a dome about 13 km (8 mi) long by 10 km (6 mi) wide. The flanks of the central intrusive mass (phonolite and trachyte) are cut by small plugs, dikes, and sills that are also of alkaline affinity (high potassium and low silica content), which include lamprophyre, syenite, nepheline syenite, and latite. These alkaline igneous rocks crop out in a northwest-trending, oval-shaped area 9 km (5.6 mi) long by 4 km (2.5 mi) wide (Staatz and others, 1979; Staatz, 1983, his plate 1). Breccia bodies are associated with the igneous intrusions, such as a heterolithic diatreme breccia near Bull Hill. Rare earth elements–bearing carbonatite dikes also intruded near the Bull Hill diatreme; these dikes are surrounded by a large zone of low-grade REE mineralization that fills thin, narrow stockwork fractures within the large alkaline intrusions. These thorium and REE deposits crop out throughout an area of about 16 km^2 (6 mi^2) in the southern Bear Lodge Mountains (Staatz, 1983).

The igneous core of the dome is microfractured and altered, thereby forming the disseminated deposits. REE and Th mineralization precipitated within thin fractures as coatings and veinlets as much as 6 mm thick. The coatings and veinlets consist predominantly of iron and manganese oxide minerals, along with potassium feldspar and quartz. The REE and thorium occupy sites in the minerals monazite, thorite, and brockite. On the basis of 52 samples collected within an area of 2.4 by 1.6 km (1.5 by 1 mi), where the alkaline rock has numerous small veinlets, Staatz and others (1979, p. 27) delineated three subareas:

> "(1) The northern area has an average grade of 0.023 percent ThO_2 and 0.75 percent combined rare-earth oxides, (2) the central area has an average grade of 0.042 percent ThO_2 and 1.71 percent combined rare- earth oxides, and (3) the southern area has an average grade of 0.03 5 percent ThO_2 and 1.35 percent combined rare-earth oxides."

They also noted that drilling "indicates that the veining extends at least 1,200 ft (365 m) below the surface."

In a broader sampling survey, Staatz (1983) collected a total of 341 samples throughout an area of 10.6 km^2 (4.1 mi^2) across the exposed core of the Bear Lodge dome (centered along Taylor Divide and Bull Hill). He found that the REE content of the disseminated deposits was about 27 times greater than their Th content. Staatz (1983, p. 1) reported, "Total rare- earth content of these samples ranged from 47 to 27,145 ppm, and the thorium content from 9.3 to 990 ppm. The amount of total rare earths of individual samples shows little correlation with that of thorium." Staatz (1983, p. 1) also stated,

> "These deposits could be mined by open pit. The Bear Lodge disseminated deposits have one of the largest resources of both total rare earths and thorium in the United States, and although the grade of both commodities is lower than some other deposits, their large size and relative cheapness of mining make them an important future resource."

Table 22. Typical Rare Earth Elements Distribution in the Bear Lodge Mountains Deposit, Wyoming

[Rare earth elements listed in order of increasing atomic number; yttrium (Y) is included with these elements because it shares chemical and physical similarities with the lanthanides. Elements listed in order of increasing atomic number. Each sample is a composite metallurgical sample. The most abundant rare earth elements in this deposit (in bold) are light rare earths elements. Source: Rare Element Resources, Ltd. (http://www.rareelementresources.com/s/Home.asp)]

Rare earth element	Oxide sample (percent)	Unoxidized sample (percent)
Lanthanum	29.3	32.5
Cerium	45.0	46.4
Praseodymium	4.8	4.3
Neodymium	16.8	13.7
Samarium	2.0	1.4
Europium	0.4	0.3
Gadolinium	0.8	0.6
Terbium	0.1	0.0
Dysprosium	0.2	0.2
Yttrium	0.5	0.5
Total	99.9	99.9

Vein deposits in the southern Bear Lodge Mountains were defined by Staatz (1983) as all tabular bodies at least 5 cm (2 in.) in thickness. Staatz (1983) mapped 26 veins in the core of the Bear Lodge uplift and described them all as thin and short—the longest vein is exposed for 137 m (450 ft). Gangue minerals are mostly potassium feldspar and quartz, with limonite, hematite, and manganese oxides. The REE and thorium in these veins is observed in monazite, brockite, and bastnasite. Staatz (1983, p. 1) reported, "Thorium content of 35 [vein] samples ranged from 0.01 to 1.2 percent, and the total rare-earth content of 21 samples from 0.23 to 9.8 percent."

Rare Element Resources, Ltd., has focused its recent REE exploration efforts (http://www.rareelementresources. com/s/ Home. asp) on Bull Mountain (near the center of the dome) and areas just to the west and southeast of Bull Mountain, where numerous carbonatite dikes are exposed. It has delineated three mineralogical zones in these deposits dependent on depth of weathering—oxide, transitional, and unoxidized (Ranta and Clark, 2010). As of 2010, they estimate that the oxide zone (the near-surface part of the "Bull Hill deposit") hosts 4.5 million tons at 4.3 percent REE oxides. They suggest that the total inferred resource of the deposit is 9.8 million tons averaging 4.1 percent REE oxides. During 2009, the company drilled five holes just to the northwest of Bull Mountain through a total of 5,141 vertical feet of veins and dikes; there, total REE oxide concentrations ranged from 2.08 to 9.12 percent. The company drilled again in the summer of 2010, which will further update and refine its REE resource estimates for this district. The rare earth elements distribution in these deposits is apparently weighted toward light REE, as shown in table 22.

Rare Element Resources has focused on the oxide portions of the Bull Hill deposit because this material has displayed favorable recovery in its metallurgical testing (Ranta and Clark, 2010). The loose, friable character of this material and fine-grained nature of the REE minerals have reportedly shown a 90 percent recovery of REE with a 13 percent REE oxide grade in the less-than-25-mm (1-in.) fraction by employing a process using simple crushing to less than 1/4 in., scrubbing, and screening (Ranta and Clark, 2010).

Phosphorite Deposits in the Southeastern United States

Location: Phosphorite deposits are found in the southeastern United States along the Atlantic Coastal Plain from North Carolina to the center of the Florida peninsula, forming a large phosphogenic province that has been subdivided into the Carolina Phosphogenic Province and the Florida Phosphogenic Province (Riggs, 1984). Phosphate deposits are also found in Virginia and Tennessee. Latitude: 27.58021 N., Longitude: 81.94569 W.; datum: WGS84

Deposit type and basic geology: Cathcart (1949) identified three main types of phosphatic rock in Florida: land- pebble phosphate, hard-rock phosphate, and river-pebble phosphate. Only the land-pebble deposits contain a substantial amount of uranium and, in some places, also contain rare earth elements. The most productive area of the land-pebble district is in Polk and Hillsborough Counties, which are located in the west-central part of the Florida peninsula and include the upper Tertiary Hawthorn Group and Bone Valley Formation. As such, mining of phosphate in Florida has been concentrated in this area (Cathcart and others, 1952). This region has been referred to more recently as the central Florida phosphate district (Van Kauwenberg and McClellan, 1990).

Status: Mining for phosphate in Florida dates back to 1883 in hard-rock deposits located near Hawthorne in Alachua County; the bulk of phosphorite is used for fertilizer. In 2009, 27.2 million metric tons (30 million tons) of phosphate rock were mined (Jasinksi, 2010), compared with 38.2 million metric tons (42.1 million tons) in 1973 (Stowasser, 1975). According to Jasinski (2010, p. 118): "In 2009, U.S. phosphate rock production and reported usage were at their lowest point since the mid-1960s, and consumption was at its lowest level since the early 1970s." The sharp decline is partly the result of a global economic crisis that started in 2008 when phosphate fertilizer producers were left with high inventories of both phosphate rock and fertilizer as farmers held out for lower prices (Jasinksi, 2010). The Atlantic Coastal Plain deposits in Florida and North Carolina account for about 85 percent of production; the majority of phosphate comes from Florida.

Production: Rare earth elements have not been extracted as a byproduct from the phosphate deposits of the southeastern United States.

Estimated resources: No one has estimated the potential resource of rare earth elements in phosphate deposits in the southeastern United States. Analyses of phosphate rock from the Bone Valley Formation indicate that the rare earth elements yttrium (Y) and lanthanum (La) are present (McKelvey and others, 1951). Samples from the Noralyn mine ranged from 0.01 to 0.1 percent Y, while those from the Bonny Lake mine varied from 0.00 1 to 0.01 percent Y. Although only one of the Bonny Lake mine samples contained La, all of the samples from the Noralyn mine contained 0.00 1 to 0.01 percent La.

Detailed Discussion

Phosphorite deposits are found in the southeastern United States along the Atlantic Coastal Plain from North Carolina to the center of the Florida peninsula, forming a large phosphogenic province that has been subdivided into the Carolina Phosphogenic Province and the Florida Phosphogenic Province (Riggs, 1984). Phosphate deposits are also found in Virginia and Tennessee. Mining for phosphate in Florida dates back to 1883 in hard-rock deposits located near Hawthorne in Alachua County (Florida Institute of

Phosphate Research (FIPR), 2010); the bulk of phosphorite is used for fertilizer. In 2009, 27.2 million metric tons (30 million tons) of phosphate rock were mined (Jasinksi, 2010), compared with 38.2 million metric tons (42.1 million tons) in 1973 (Stowasser, 1975). According to Jasinski (2010, p. 118): "In 2009, U.S. phosphate rock production and reported usage were at their lowest point since the mid-1960s, and consumption was at its lowest level since the early 1970s." The sharp decline is partly the result of a global economic crisis that started in 2008 when phosphate fertilizer producers were left with high inventories of both phosphate rock and fertilizer as farmers held out for lower prices (Jasinksi, 2010). In addition, the richest orebodies in Florida have been depleted, resulting in massive consolidation of phosphate mining companies in the State. The Atlantic Coastal Plain deposits in Florida and North Carolina account for about 85 percent of production, and the majority of phosphate comes from Florida.

Cathcart (1949) identified three main types of phosphatic rock in Florida: land-pebble phosphate, hard-rock phosphate, and river-pebble phosphate. Only the land-pebble deposits contain a substantial amount of uranium and, in some places, also contain rare earth elements (REE). Historically, the most productive area of the land-pebble district occurs in Polk and Hillsborough Counties, which are located in the west-central part of the Florida peninsula, and which contain the Hawthorn Group and Bone Valley Formation. As such, mining of phosphate in Florida, which began in 1888, has been concentrated in this area (Cathcart and others, 1952). This region has been referred to more recently as the central Florida phosphate district (Van Kauwenberg and McClellan, 1990). As time passed, the mining became more concentrated in the Four Corners Area (southwest of Orlando), as operations moved down dip to the south and west.

The central Florida platform, or phosphate district, is a shield-shaped area (figure 17) that is about 7,252 km^2 (2,800 mi^2) and includes Polk, Hillsborough, Hardee, Manatee, Sarasota, DeSoto, and Highland Counties (Cathcart, 1949). The middle Miocene Hawthorn Group is found throughout the entire Florida peninsula, except within the Ocala High (or "Ocala Upland") and the Sanford High. The Hawthorn Group contains considerable phosphorite (Riggs, 1980). A preliminary investigation report by Cathcart and others (1952) divides the Hawthorn Group into three lithologic units: a lowermost unit containing dolomite, phosphatic marls, limestone, and clay and quartz sands; a middle series of interbedded phosphatic sands, clays, and sandy clays; and an uppermost unit consisting of dolomites and dolomitic limestone. In general, the phosphorite grains range in size from less than 0.1 mm to 10 cm (0.0039 to 3.937 in.) in diameter.

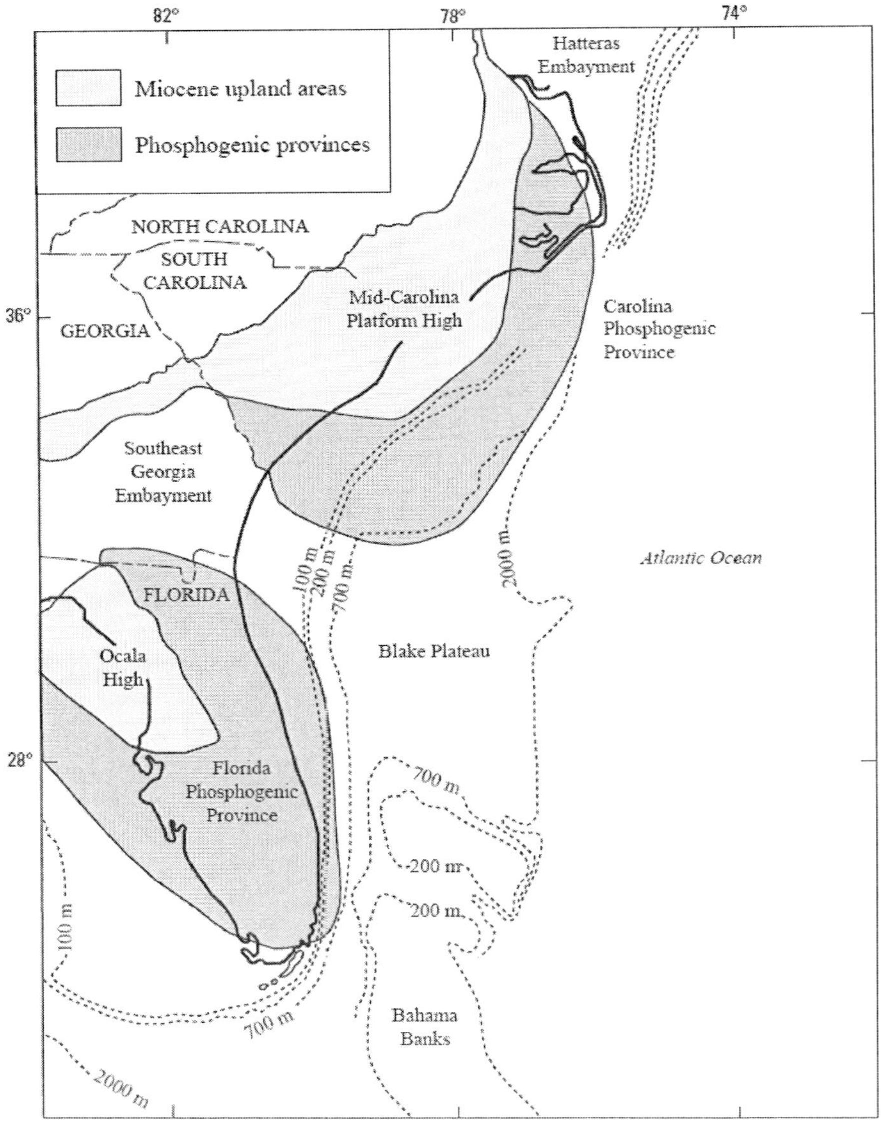

Figure 17. Phosphogenic provinces of the southeastern United States. Modified from Van Kauwenbergh and McClellan (1990).

Uncomformably overlying the Hawthorn Group, the Bone Valley Formation contains phosphate that is typically a higher grade, most likely the result of mechanical reworking and subsequent precipitation during the Pliocene (Cathcart, 1949). As a result, the Bone Valley Formation is a lithochemical sequence in which the phosphorite appears sporadically in irregular concentration and distribution (Riggs, 1980). The formation consists of a lower zone of poorly sorted clay, silt, sand, and gravel (Cathcart and others, 1952). Nearly all the gravel in this unit, as well as a large part of the sand and clay, is phosphate, specifically carbonate-fluorapatite (Cathcart and others, 1952; Altschuler and others, 1952). The rest of the sand consists of quartz, with minor amounts of feldspar and trace amounts of ilmenite, zircon, tourmaline, staurolite, and other minerals. The clay minerals include palygorskite, montmorillonite, and kaolinite. The minerals in the upper zone of the Bone Valley Formation are quartz sand, wavellite and pseudowavellite (aluminum phosphates), crandallite, and kaolinite.

Uranium (U) in the Bone Valley Formation has a maximum concentration of 0.02 percent uranium in the basal section and an average concentration of about 0.01 percent uranium (Cathcart, 1949). In the Hawthorn Group, there is a low content of uranium in fresh, unweathered material. However, after leaching, the phosphate-rich rocks of the Hawthorn Group yield a maximum of 0.01 percent U. In Manatee and Hardee Counties, which are south of the high-grade district, only minor amounts of uranium occur in the Bone Valley, Hawthorn, and overlying Pleistocene sand formations.

In addition to uranium, semiquantitative spectrographic analyses on phosphate rock from the Bone Valley Formation indicate that the rare earth elements yttrium (Y) and lanthanum (La) are present (McKelvey and others, 1951). Samples from the decommissioned Noralyn mine range from 0.01 to 0.1 percent Y while those from the decommissioned Bonny Lake mine range from 0.001 to 0.01 percent Y. Although only one of the Bonny Lake mine samples contains La, all of the samples from the Noralyn mine contain 0.001 to 0.01 percent La.

Several metals are also present in nearly all samples from the Bone Valley Formation such as vanadium, manganese, chromium, barium, strontium, and titanium, with concentrations that range from 0.01 to 0.1 percent (McKelvey and others, 1951). Nickel, copper, zirconium, and lead are also present in amounts greater than 0.00 1 percent. The Bonny Lake mine samples contain 0.00 1 to 0.01 percent tin whereas those from the Noralyn mine contain 0.000 1 to 0.00 1 percent. McKelvey and others (1951) also report gallium in all samples from the two mines.

Although the distribution of the metals is unknown, the REE, U, Ba, and Sr are most likely located in the carbonatefluorapatite (francolite) lattice or absorbed on its surface (Cathcart and others, 1952; Clarke and Altschuler, 1958). In addition, because of the small quantities of the various elements, recovery may be possible only when ore is taken into acid solution. More recent research regarding the rare earth geochemistry of phosphorites suggests that the two main factors influencing REE concentrations in marine sedimentary phosphates are primary compositional differences and postdepositional processes such as diagenesis, the removal of mineral species, and reworking of deposits by winnowing action (Van Kauwenbergh and McClellan, 1990). While the North Carolina phosphorites appear to be unaltered, the Florida phosphorites have undergone postdepositional alteration through the removal of carbonate mineral species, systematic decarbonatization of francolite, deposition of iron and aluminum phosphates and of clay minerals in alteration profiles.

MacArthur and Walsh (1984) determined that the abundance of REE in francolite reflects the REE source and mechanism of incorporation. Zanin and Zamirailova (2009) suggested that the REE concentrations of supergene phosphorites are the result of weathering, the physicochemical conditions of phosphorite formation, presence of a biogenic component, and structural type of the phosphorites. In particular, phosphorite from the weathering zone of sedimentary rocks, such as those that are found in Florida and Tennessee, yield an average total REE content of 27 ppm (Zanin and Zamirailova, 2009).

Further research on the REE and trace metal contents of Florida phosphorites is critical to more fully assessing its viability as an economic resource. Additional REE and trace metal work on the North Carolina phosphorites is also warranted, as little has been reported at this point.

Placer Rare Earth Elements Deposits

Placer deposits are sand, silt, and cobble-size sediments deposited in streams, rivers, and beaches, also referred to as "alluvial" deposits. Globally and in the United States, alluvial accumulations of monazite are a valuable type of rare earth elements (REE)–thorium (Th) deposit. For example, the alluvial monazite deposits of the Coastal Belt of southernmost India are thought to represent one of the largest thorium resources in the world. The monazite deposits of southern India contain detrital heavy minerals and are found in piedmont lakes, shallow seas, parts of the beaches (figure 18), sand bars across the mouth of rivers, deltas, and sand dunes behind the beaches (Bhola and others, 1958). A study by Mahadevan and others (1958) estimated that the beach sands of the southwestern coast of India alone contain estimated reserves of 446,400 metric tons (492,200 tons) of monazite, in which the ThO_2 content of the monazite ranges from 7.5 to 9 percent.

Monazite's resistance to chemical weathering and high specific gravity account for its association in alluvial (placer) deposits with other resistant heavy minerals such as ilmenite, magnetite, rutile, and zircon. Monazite weathers from alkaline crystalline rocks of the surrounding region and is transported downstream and deposited by alluvial processes.

In the United States, alluvial deposits of monazite are known in the Carolina Piedmont of North and South Carolina, the beach deposits of northeastern Florida through southeastern Georgia, and the intermontane valleys of Idaho. In the past, these unconsolidated stream deposits were mined by small-scale sluices (Idaho and North and South Carolina) and dredges (Idaho); beach sands were mined by large shovels (Florida and Georgia). Sluicing instantly produces a heavy- mineral separate, but even in the case of the dredge or shovel operations, the heavy-mineral separation was still performed at the site. Thus, the mining of REE and thorium from alluvial deposits has the advantages of relative ease of mining and rapid mineral separation, in contrast to hard-rock mining. Another benefit of placer deposits is the potential for coproducts. Coproducts can include REE and thorium obtained from monazite; titanium from ilmenite and rutile; iron from magnetite; zirconium and hafnium from zircon; and industrial-grade garnet, staurolite, tourmaline, kyanite, and sillimanite, which are used as abrasives and refractory minerals.

The three monazite placer districts highlighted in this report—North and South Carolina stream deposits, Florida- Georgia beaches, and Idaho stream deposits—are the largest volume alluvial REE-thorium deposits known in the United States. The geology and estimated monazite resources of these districts have been well described by Staatz and others (1979, the North and South Carolina placer deposits, p. 33–39) and Staatz and others (1980, the Idaho stream placers, p. 9–18, and the Florida beach deposits, p. 3–9); numerous references cited therein provide more detailed information on these deposits. Thus, here we only briefly summarize the findings of these earlier studies.

IDAHO—PLACER DEPOSITS

Location: At least 11 monazite-bearing placer districts exist in the valleys of a region extending north of Boise, Idaho, and along the western flank of the Idaho batholith. Latitude: 44.40416 N., Longitude: 115.35356 W.; datum: WGS84

Deposit type and basic geology: Monazite-bearing alluvial stream deposits (placers) exist in the valleys of a region that extends north of Boise, Idaho along the western flank of the Idaho batholith. The primary source of the resistant rare earth elements (REE)–thorium-bearing minerals in the Idaho placers is thought to be the Idaho batholith, in particular the quartz monzonite and pegmatite phases of the batholith (Mackin and Schmidt, 1957). The most common heavy minerals in the alluvial deposits (in generally decreasing amounts) are ilmenite, magnetite, sphene, garnet, monazite, euxenite, zircon, and uranothorite (uranium-rich thorite). In addition to REE and thorium from monazite and euxenite, the Idaho placer deposits (and the abandoned historic dredge waste piles) contain coproducts of titanium (in ilmenite), and niobium and tantalum (in euxenite).

Status: Currently (2010), there is no active exploration in this district.

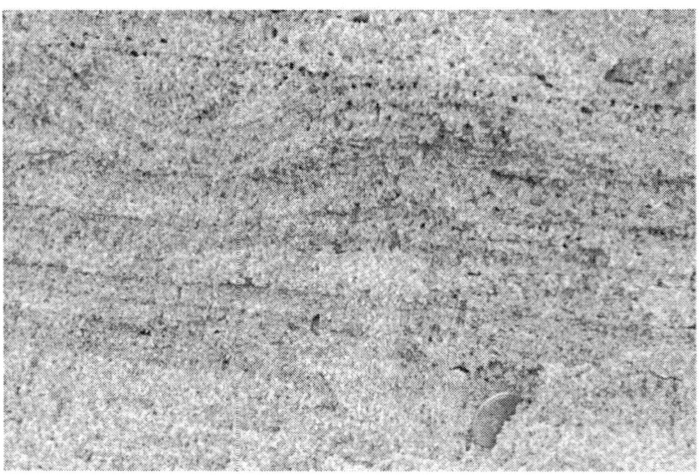

Figure 18. Heavy-mineral layers ("black sand") in a quartz beach sand, Chennai, India. A penny provides a scale. (Photograph by Mark A. Wilson, Department of Geology, The College of Wooster, Wooster, Ohio; used with permission).

Production: In the 1950s, two areas of west-central Idaho were mined by dredges for monazite, Long Valley and Bear Valley. Beginning in September 1950, Long Valley was worked by three dredges that had earlier been used to recover gold but later were converted (with assistance from the U.S. Bureau of Mines under the sponsorship of the U.S. Atomic Energy Commission) to recover monazite. The history of these dredging operations is described by Argall (1954) and Staatz and others (1980, p. 9–16, and references cited therein). During this 5-year period, Staatz and others (1980) estimated that the three dredges recovered 6,430 metric tons (7,085 tons) of monazite containing 269 metric tons (297 tons) of thorium oxide. Dredging ended here in mid-1955, when the government stockpile order was fulfilled. The Bear Valley placers were worked by first one dredge in 1955, then a second in 1956, with the intent to recover Nb and Ta for another Federal government contract. According to Staatz and others (1980, p. 10), "from alluvium of Bear Valley, 2,049 short tons [1,858 metric tons] of euxenite, 83.5 tons [75.7 metric tons] of columbite, and 54,862 tons [49,760 metric tons] of ilmenite were recovered." No records of monazite recovery were kept.

Estimated resources: U.S. Geological Survey geologists (Staatz and others, 1980) extensively reviewed results of the 1 950s government dredging program in central Idaho. They determined that the five most important monazite districts are Long Valley, Bear Valley, the Boise Basin, the Burgdorf-Warren area, and the Elk City–Newsome area. The reported thorium oxide contents of monazite in the Idaho placer deposits range from 2.2 to 6.24 percent. The few analyses of REE in monazites from Idaho placers indicate that these monazites contain 63 percent total REE oxides Staatz and others (1980). Staatz and others (1980) calculated thorium reserves for each of the five major placer districts individually; in total, the five districts have total reserves of about 9,130 metric tons (10,060 tons) of thorium oxide. The REE resources of the five placer districts would presumably be at least ten times the thorium resource, because the typical monazite contains about 63 percent total REE oxides and 2.2–6.24 percent thorium oxide.

Detailed Discussion

At least 11 monazite-bearing placer districts exist in the valleys of a region extending north of Boise, Idaho, and along the western flank of the Idaho batholith (figure 19). Monazite was first recognized here in 1896 as the heavy, yellow to brownish-yellow mineral that collected with other heavy minerals and gold within the sluice boxes of gold placer operations in the Boise Basin near Idaho City, Centerville, and Placerville (Lindgren, 1897). In 1909, a mill designed to capture the monazite was built by the Centerville Mining and Milling Co. Only a small amount of monazite concentrate was produced for its thorium content before the mill burned down in a forest fire in 1910.

In the 1950s, two areas of west-central Idaho were mined by dredges for monazite recovery, Long Valley and Bear Valley (figures. 19 and 20). Beginning in September 1950, Long Valley was worked by three dredges that were earlier used to recover gold but later were converted (with assistance from the U.S. Bureau of Mines under the sponsorship of the U.S. Atomic Energy Commission) to recover monazite. The history of these dredging operations is described by Argall (1954) and Staatz and others (1980, p. 9–16, and references cited therein). The heavy minerals recovered in the Long Valley district were dominated by ilmenite (84 percent of heavy minerals), followed by monazite (8 percent), garnet (5 percent), and zircon (3 percent). During this 5-year period, Staatz and others (1980) estimated that the three dredges recovered 6,430 metric tons (7,085 tons) of monazite containing 269 metric tons (297 tons) of thorium oxide. The dredging ended here in mid-1955, when the government stockpile order was fulfilled.

Rare earth elements (REE) and thorium were also unintentionally recovered within the minerals euxenite and monazite from the Bear Valley placers. The Bear Valley placers were worked by first one dredge in 1955, then a second in 1956, with the intent to recover Nb and Ta for another Federal government contract. According to Staatz and others (1980, p. 10), "from alluvium of Bear Valley, 2,049 short tons [1,858 metric tons] of euxenite, 83.5 tons [75.7 metric tons] of columbite, and 54,862 tons [49,760 metric tons] of ilmenite were recovered." No records of the monazite recovery were kept.

Most of the Idaho monazite resource data used by Staatz and others (1980) came from a 1950s program in which the U.S. Atomic Energy Commission funded the U.S. Bureau of Mines to identify new domestic monazite reserves (Storch and Holt, 1963). The USGS (Staatz and others, 1980) extensively reviewed results of this program, as attested by U.S. Bureau of Mines literature, interviews with former dredge companies, and field reconnaissance and sampling of the major monazite placer districts in Idaho. They determined that the five most important monazite districts are Long Valley, Bear Valley, the Boise Basin, the Burgdorf-Warren area, and the Elk City–Newsome area. The reported thorium oxide contents of monazite in the Idaho placer deposits range from 2.2 to 6.24 percent. The few analyses of REE in monazites from Idaho placers indicated that these monazites contain 63 percent total REE oxides Staatz and others (1980). Staatz and others (1980) calculated thorium reserves for each of the five major placer districts individually; in total, the five districts have total reserves of about 9,130 metric tons (10,060 tons) of thorium oxide. The REE resources of the five placer districts would presumably be at least 10 times the thorium resource, because the typical monazite contains about 63 percent total REE oxides and 2.2–6.24 percent thorium oxide.

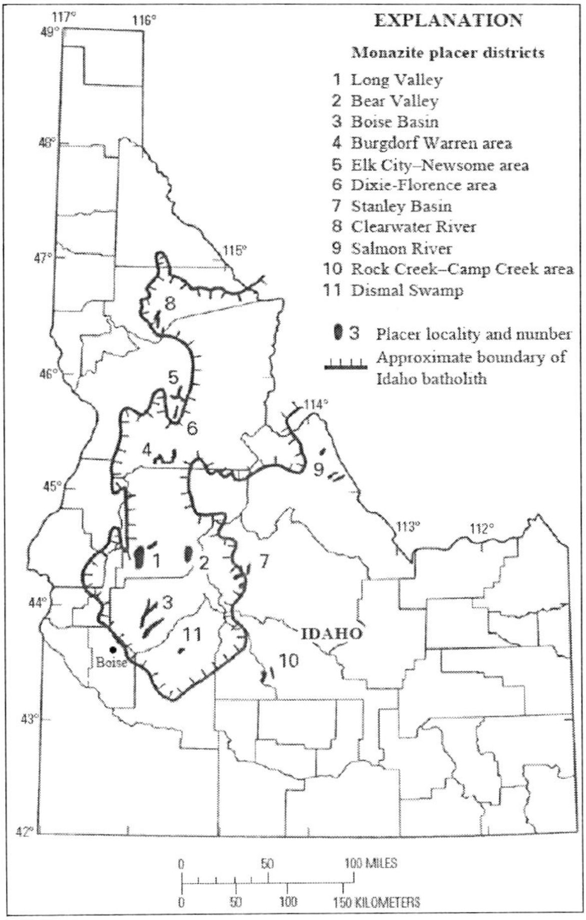

Figure 19. Generalized map of known monazite placer districts in Idaho. From Staatz and others (1980).

Figure 20. Porter Brothers dredge in Bear Valley, Idaho, worked columbium (now called niobium)-tantalum placer deposits. This dredge operated from 1955 to 1960; exact date of photograph is unknown. Monazite was produced as a byproduct from these placer deposits. (Photograph from Savage (1961, figure 13); used with permission of Idaho Geological Survey).

The primary source of the resistant REE-thorium– bearing minerals in the Idaho placers is thought to be the Idaho batholith, in particular the quartz monzonite and pegmatite phases of the batholith (Mackin and Schmidt, 1957). The most common heavy minerals in the alluvial deposits (in generally decreasing amounts) are ilmenite, magnetite, sphene, garnet, monazite, euxenite, zircon, and uranothorite (uranium- rich thorite). In addition to REE and thorium from monazite and euxenite, the Idaho placer deposits (and the abandoned historic dredge waste piles) contain coproducts of titanium (in ilmenite), and niobium and tantalum (in euxenite).

North and South Carolina—Placer Deposits

Location: The high-grade monazite placers of the Piedmont of North Carolina and South Carolina are deposited between the Catawba River in the northeast and the Savannah River in the southwest, along a belt that extends from east- central Virginia southwestward into Alabama. Latitude: 33.40277 N., Longitude: 81.79004 W.; datum: WGS84

Deposit type and basic geology: Monazite-bearing alluvial stream deposits (placers) exist in the valleys of the Piedmont of North Carolina and South Carolina. The stream- sediment deposits in this region are generally consistent in character; the heavy-mineral concentrations are greatest in the headwaters areas. The alluvium is deposited in flat valleys, forming well-bedded, poorly graded layers of unconsolidated sediment. Stacked layers contain gravel, sand, clay, and clayey silt, at an average total thickness of about 4.5 m (15 ft) (Staatz and others, 1979). Monazite typically is found in all units, but it is generally most abundant in the basal gravel layers and least abundant in the clay layers. According to Staatz and others (1979), the heavy-mineral content of the placer deposits of the Piedmont region ranges from 0.15 to 2.0 percent; monazite makes up about 3.5–13 percent of the heavy minerals. Other parts of the heavy-mineral fraction include ilmenite, 20–70 percent; garnet, 2–50 percent; rutile, 0.3–7 percent; zircon, trace to 14 percent; and sillimanite and kyanite together, trace to 20 percent. In some placers, additional heavy minerals include epidote, magnetite, xenotime, tourmaline, sphene, staurolite, andalusite, and an unidentified black radioactive mineral (Staatz and others, 1979). Analysis of 52 samples of alluvial monazite from this region (Mertie, 1975) found that the monazite contains 60–63 percent total rare earth elements (REE) oxides and 2.5–7.8 percent thorium oxide content, with a mean value of 5.67 percent thorium.

Status: Currently (2010), there is no reported exploration in this region.

Production: In 1887, a few short tons of monazite were produced from stream deposits in the Piedmont region of North and South Carolina, giving this region the distinction of being the world's first supplier of thorium (Olson and Overstreet, 1964). Monazite-bearing placers of this region were worked by small-scale sluice operations from 1887 to 1911 and 1915 to 1917; they produced a total of 4,973 metric tons (5,483 tons) of monazite (Overstreet and others, 1968). Monazite mining ended here in 1917 because beach deposits in India and Brazil were producing thorium at lower cost.

Estimated resources: For the 13 largest placer deposits of the Piedmont region of North and South Carolina, Staatz and others (1979) estimated total reserves of about 4,800 metric tons (5,300 tons) of thorium oxide and potential thorium resources seven times as high. Using mean concentrations of REE oxides (60–63 percent) and thorium oxide (5.67 percent) found in monazite from these placer deposits, a REE resource of roughly 53,000 metric tons (58,400 tons) of REE oxide is suggested.

Detailed Discussion

In 1887, a few short tons of monazite were produced from stream deposits in the Piedmont region of North and South Carolina, giving this region the distinction of being the world's first supplier of thorium (Olson and Overstreet, 1964). Monazite-bearing placers of this region were worked by small-scale sluice operations from 1887 to 1911 and 1915 to 1917; they produced a total of 4,973 metric tons (5,483 tons) of monazite (Overstreet and others, 1968). Monazite mining ended here in 1917, not because reserves had been exhausted but rather because the beach deposits of India and Brazil were producing thorium at lower cost.

The high-grade monazite placers of the Piedmont of North Carolina and South Carolina are deposited between the Catawba River in the northeast and the Savannah River in the southwest (figure 21), along a belt that extends from east-central Virginia southwestward into Alabama (Mertie, 1975). The stream-sediment deposits in this region are generally consistent in character; the heavy-mineral concentrations are greatest in the headwaters areas. The alluvium is deposited in flat valleys, forming well-bedded, poorly graded layers of unconsolidated sediment. Stacked layers contain gravel, sand, clay, and clayey silt, at an average total thickness of about 4.5 m (15 ft) (Staatz and others, 1979). Monazite typically is found in all units, but it is generally most abundant in the basal gravel layers and least abundant in the clay layers.

The Piedmont region is underlain by crystalline, high- grade metamorphic rocks intruded by quartz monzonite and pegmatite. The monzonite and pegmatite intrusions may be monazite bearing or monazite free. Overstreet (1967) suggested that the primary source of the alluvial monazite was the high-grade metamorphic rocks, particularly sillimanite schist. Other metamorphic rocks in the area include mica and hornblende gneisses, amphibolites, and additional varieties of schist (Mertie, 1975). Other igneous country rocks include diorite, rhyolite with associated pyroclastic rocks, gabbro, diabase, ultrabasic rocks, and several kinds of granite, such as monzonite, quartz monzonite, and granodiorite.

Various rivers carried monazite and other heavy minerals eastward from the Piedmont region, such that all the regions in the Coastal Plain may have also received various amounts of heavy minerals (Staatz and others, 1979). In particular, the Late Cretaceous Tuscaloosa Formation received a large amount of monazite, albeit widely dispersed. Although this area has not been as well explored as the Piedmont, it is known that the Tuscaloosa directly overlaps the crystalline rocks of the Piedmont and that streams in the area have reworked the Tuscaloosa sand so that in places heavy-mineral placers containing monazite have been identified (Staatz and others, 1979). The best-known placer is Horse Creek, which is southwest of Aiken, South Carolina, and was the site of the first large-scale mining of stream placers for monazite and other heavy minerals in the Carolinas. Dredging in this area between the summers of 1955 and 1958 (Williams, 1967) found heavy-mineral contents of about 1−1.5 percent, of which fraction monazite formed about 8 percent (Mertie, 1975). Overall, these dredging operations recovered monazite, ilmenite, rutile, zircon, and staurolite (Williams, 1967).

According to Staatz and others (1979), the heavy-mineral content of the placer deposits of the Piedmont region ranges from 0.15 to 2.0 percent, and monazite forms about 3.5–13 percent of the heavy minerals. Other parts of the heavy- mineral fraction contain ilmenite, 20–70 percent; garnet, 2–50 percent; rutile, 0.3–7 percent; zircon, trace to 14 percent; and sillimanite and kyanite together, trace to 20 percent. In some placers, additional heavy minerals include epidote, magnetite, xenotime, tourmaline, sphene, staurolite, andalusite, and an unidentified black radioactive mineral (Staatz and others, 1979). Analysis of 52 samples of alluvial monazite from this region (Mertie, 1975) found that the monazite contains 60–63 percent total REE oxides and 2.5–7.8 percent (mean, 5.67 percent) thorium oxide.

The Fall Line is the sharp topographic break that marks the boundary between the Piedmont and the Coastal Plain (figure 21). East of the Fall Line, the heavy-mineral distribution in two deposits (the only ones for which data are available) shows several differences from that of the Piedmont. First, the abundance of staurolite increases to 7 and 38 percent of the two deposits (Kline and others, 1954, p. 27; Mertie, 1975, p. 27). Rutile and zircon are also more abundant in these deposits than in the Piedmont placers; rutile makes up 15 and 10 percent of the heavy-mineral content and zircon 19 and 11 percent. However, monazite concentrations are similar to those in placers in the Piedmont.

For the 13 largest placer deposits of the Piedmont region of North and South Carolina, Staatz and others (1979) estimated total reserves of about 4,800 metric tons (5,300 tons) of thorium oxide, with potential thorium resources seven times as high. However, using the mean concentrations of REE oxides (60–63 percent) and thorium oxide (5.67 percent) found in monazite from these placer deposits, a REE resource of roughly 53,000 metric tons (58,400 tons) of REE oxide is suggested. The estimate of Staatz and others (1979) was based on regional monazite resource studies by Overstreet and others (1959) and Overstreet (1967), and on studies of individual drainage basins by several others (see Staatz and others, 1979, p. 37). Future exploration for monazite placer deposits in the Mid-Atlantic region can be aided by regional stream-sediment geochemistry datasets assembled by the USGS (Grosz, 1993).

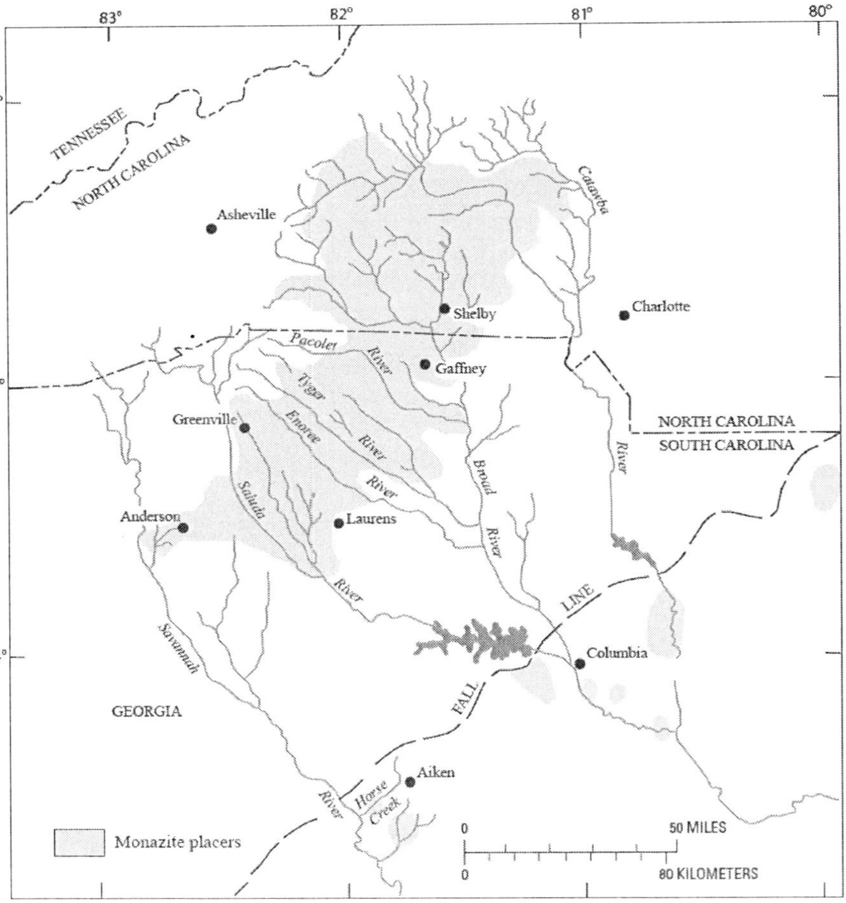

Figure 21. Map of monazite placers of North Carolina and South Carolina. Modified from Staatz and others (1979).

FLORIDA-GEORGIA—BEACH PLACER DEPOSITS

Location: Monazite accumulations are situated in the modern and raised Pleistocene and Pliocene beach deposits of northeastern Florida and southeastern Georgia. Latitude: 29.85997 N., Longitude: 8 1.73711 W.; datum: WGS84

Deposit type and basic geology: The modern and raised Pleistocene and Pliocene beach deposits of northeastern Florida and southeastern Georgia host low-grade but persistent concentrations of monazite. Heavy minerals constitute a small part of the modern beach sands, and monazite forms a small part of the heavy minerals. However, because of the large tonnage of suitable beach-sand deposits and the ease of mining and processing this material, these beach deposits represent a potential rare earth elements (REE) and thorium resource. The most abundant heavy mineral in the southeastern United States shoreline deposits is ilmenite, which in many places forms more than 50 percent of the heavy-mineral fraction; monazite forms a minor part of the heavy-mineral fraction, usually less than 1 percent (Staatz and others, 1980). Monazite from the Florida beach placers contains about 50–60 percent total REE oxides (Kremers, 1958; Staatz and others, 1980) and 4–5 percent thorium oxide (Calver, 1957). Small amounts of uranium, averaging 0.55 percent, are also present in the monazite (Calver, 1957).

Status: Currently (2010), there is no reported exploration in this district.

Production: Some of the former heavy-mineral operations in the Pleistocene beach deposits of Florida were once domestic suppliers of monazite. Staatz and others (1980, p. 3) reported, "During 1978 monazite was produced from two of the three operating heavy-mineral deposits in Florida: Titanium Enterprises at Green Cove Springs and Humphrey Mining Corp. at Boulogne recovered monazite as a byproduct." Trace amounts of monazite were also mined from the large Trail Ridge orebody south of Jacksonville in 1949 by E.I. du Pont de Nemours and Company. Overall, these deposits were mined primarily for titanium (in ilmenite and rutile), which was used by the pigment industry. Other minerals sold from the three deposits include kyanite, sillimanite, staurolite, garnet, zircon, and the host sand itself. Mining ceased in this area in late 1978 because of increasing environmental regulations that made mining operations more costly.

Estimated resources: Staatz and others (1980) estimated that the beach placer deposits of this region contain total reserves of about 198,000 metric tons (218,000 tons) of rare earth elements oxides, 14,700 metric tons (16,200 short tons) of thorium oxide, and 1,490 metric tons (1,640 tons) of uranium oxide, all of which are hosted in 330,000 metric tons (364,000 tons) of monazite.

Detailed Discussion

The modern and raised Pleistocene and Pliocene beach deposits of northeastern Florida and southeastern Georgia host low-grade but persistent concentrations of monazite (figures. 22 and 23). Heavy minerals constitute a small part of the modern beach sands, and monazite forms a small part of the heavy minerals. However, because of the large tonnage of suitable beach sand deposits and the ease of mining and processing this material, these beach deposits represent a potential rare earth elements (REE) and thorium resource. Staatz and others (1980) estimated that the beach placer deposits of this region contain total reserves of about 198,000 metric tons (218,000 tons) of REE oxides, 14,700 metric tons (16,200 short tons) of thorium oxide, and 1,490 metric tons (1,640 tons) of uranium oxide, all of which are hosted in 330,000 metric tons (364,000 tons) of monazite.

Some of the heavy-mineral operations in the Pleistocene beach deposits of Florida were once domestic suppliers of monazite (figure 23). Staatz and others (1980, p. 3) noted, "During 1978 monazite was produced from two of the three operating heavy-mineral deposits in Florida: Titanium Enterprises at Green Cove Springs and Humphrey Mining Corp. at Boulogne recovered monazite as a byproduct." Trace amounts of monazite were also mined from the large Trail Ridge orebody south of Jacksonville in 1949 by E.I. du Pont de Nemours and Company. The Boulogne orebody underlies an area of about 4 km^2 (1.5 mi^2) and was mined to a depth of 4.6 m (15 ft), whereas the Trail Ridge deposit has been traced for about 29 km (18 mi) (Staatz and others, 1980).

Overall, these deposits were mined primarily for titanium (in ilmenite and rutile), which was used by the pigment industry. Other minerals sold from the three deposits include kyanite, sillimanite, staurolite, garnet, zircon, and the host sand itself. Mining ceased in this area in late 1978 because of increasing environmental regulations that made mining operations more costly. Steep increases in coastal real-estate values also influenced the situation, such that the heavy-mineral concentrations in the modern beaches became much less valuable than the real estate they occupy. Nonetheless, while the modern beaches hosted most of the early heavy-mineral operations, the older inland beaches are larger, have more uniform distribution of heavy minerals, and are not as vulnerable to severe storms, which made them favorable before the real estate market skyrocketed (Overstreet, 1967, p. 125). Staatz and others (1980, p. 3–4, and references cited therein) describe the mining history of the heavy-mineral beach placers of the northeastern Florida and southeastern Georgia area in greater detail.

In general, the monazite-bearing sands in the raised Pleistocene and Pliocene beach deposits lie as much as 80 km (50 mi) inland, making them deposits of former shorelines. These relict shorelines, which lie 3–33 m (10–108 ft) above the sea level, have been noted in the outer coastal plain region from Maryland to Florida. Once referred to as "marine terraces," the relict shorelines were divided into different levels on the basis of elevation and tectonic stability and then correlated with interglacial stages (MacNeil, 1950; Doering, 1960). Further investigation revealed that the relict shorelines more closely resembled barrier islands, suggesting that the coastal plain was warped during the Pleistocene (Winkler and Howard, 1977). Because monazite contains radioactive thorium, future monazite exploration in the eastern United States coastal plain areas can benefit from several aeroradiometric maps that were compiled and interpreted by the USGS (Force and others, 1982; Grosz, 1983; Grosz and others, 1989; Owens and others, 1989).

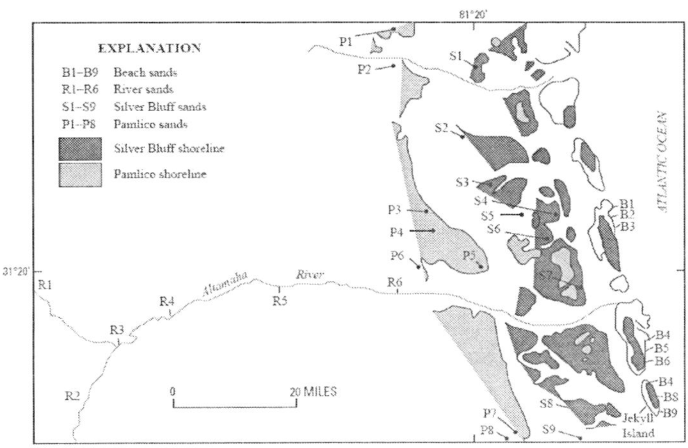

Figure 22. Map of Recent and Pleistocene sands of Georgia. Modified from Neiheisel (1962).

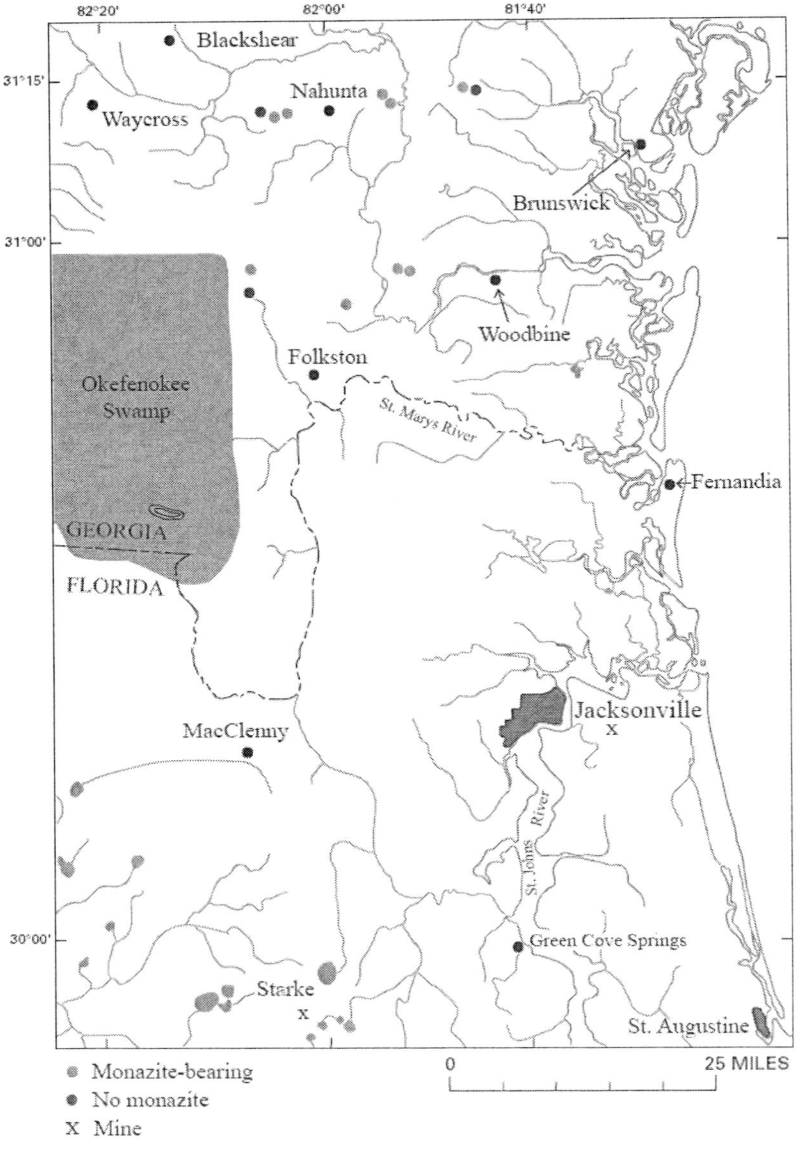

Figure 23. Sample and mine locations of monazite in southeastern Georgia and northeastern Florida. Modified from Mertie (1975).

The monazite and associated heavy minerals in the relict shoreline deposits of the southeastern United States were eroded from crystalline rocks of the Piedmont province (Mertie, 1953), carried towards the Atlantic Ocean by streams and rivers, and eventually redeposited by coastal processes. The natural concentration of heavy minerals in the shoreline area is a multistage process that involves transport by longshore drift; gravity separation by specific gravity, particle size, and shape; differential chemical weathering (Neiheisel, 1962); wave action; and, in some parts of the coastal environment, the actions of tides. All of these forces rework the sediments in the shoreline environments through time and naturally concentrate the heavy minerals (see Force, 1991, p. 73–84).

Although the most abundant heavy mineral in the south-eastern United States shoreline deposits is ilmenite, which in many places composes more than 50 percent of the heavy- mineral fraction, monazite usually forms less than 1 percent of the heavy-mineral fraction (Staatz and others, 1980). Zircon generally composes 10–20 percent of the heavy minerals. Monazite from the Florida beach placers contains about 50–60 percent total REE oxides (Kremers, 1958; Staatz and others, 1980) and 4–5 percent thorium oxide (Calver, 1957). Small amounts of uranium are also present in the monazite, averaging 0.55 percent (Calver, 1957).

Despite the low concentrations of monazite (and thus, REE and thorium) in the typical coastal placer deposit of the southeastern United States, these deposits have three distinct advantages as potential sources of REE and thorium: they are relatively easy to excavate; it is relatively easy to separate the heavy-mineral fraction onsite; and they contain several salable mineral products. Mining in both beach placers and on relict shorelines is possible using open-pit methods, and overburden is rarely greater than 4–5 m (13–16 ft) thick. In addition, ground¬water in Florida and Georgia is shallow enough that monazite and other heavy minerals can be mined by a dredge floating on a pond (Staatz and others, 1980). However, at present the Bou¬lougne and Jacksonville deposits are essentially mined out.

REFERENCES CITED

[1] Altschuler, Z. S., Berman, Sol & Cuttitta, Frank. (1966). *Rare earths in phosphorite—Geochemistry and potential recovery*: U.S. Geological Survey Open-File Report 66–3, 35 p.

[2] Castor, S. B. (2008). Rare earth deposits of North America: *Resource Geology*, v. *58*, no. 4, 337–347.

[3] Castor, S. B. & Hedrick, J. B. (2006). Rare earth elements, *in* Kogel,

J.E., Trivedi, N.C., Barker, J.M., and Krukowski, S.T., eds., Industrial minerals & rocks—Commodities, markets, and uses (7th ed.): Littleton, Colo., Society for Mining, Metallurgy, and Exploration, Inc. (SME), 769–792.

[4] Emsley, John. (2001). *Nature's building blocks—An A–Z guide to the elements: Oxford,* Oxford University Press, 538 p.

[5] Gupta, C. K. & Krishnamurthy, N. (2005). *Extractive metallurgy of rare earths: Boca Raton*, Florida, CRC Press, 484 p.

[6] Hedrick, James B. (2004). Rare earths, *in Minerals Yearbook*, v. 1, Metals and Minerals: U.S. Geological Survey, 60.1–60.15.

[7] Heymann, Les. (2010). *The valuation of rare earth deposits: Paper presented at the Society for Mining, Metallurgy, and Exploration Annual Meeting*, Phoenix, Arizona, March, 2010.

[8] Jackson, W. D. & Christiansen, G. (1993). *International Strategic Minerals Inventory summary report—Rare earth oxides*: U.S. Geological Survey Circular 930–N, 68 p.

[9] Lide, D. R. (1997). Abundance of elements in the earth's crust and sea, *in CRC handbook of physics and chemistry,* 78th edition: Boca Raton, Florida, CRC Press, p. 14.

[10] Lynas Corporation Ltd. (2010). Will there be sufficient rare earths to meet demand from clean energy technology?: Presentation at International Minor Metals Conference, London, April, 2010, accessed July 5, 2010, at *www.* lynascorp.com/content/upload/files/ Presentations/MMTA_ *APRIL_2010.pdf.*

[11] Maksimović, Z. J. & Pantó, Gy. (1996). Authigenic rare earth minerals in karst-bauxite and karstic nickel deposits, *in* Jones, A.P., Wall, Frances, and Williams, Terry, eds., Rare earth minerals—Chemistry, origin and ore deposits: London, Chapman and Hall, p. 257–279.

[12] Mason, B. & Moore, C. B. (1982). *Principles of geochemistry*, 4th ed.: New York, Wiley, 344 p.

[13] McGill, I. (1997). Rare earth metals, *in* Habashi, F., ed., Handbook of extractive metallurgy: *Weinheim, New York, WileyVCH,* v. *3*, 1695–1741.

[14] O'Driscoll, M. (2003). Regeneration in autocatalysts: *Industrial Minerals,* v. *424*, 21–54.

[15] Ren, X. (1985). A brief account of rare earth mineralization in China, *in* Xu, Q., ed., New frontiers in REE science and applications: Beijing, China, Science Press, 39–41.

[16] Sabot, J. L. & Maestro, P. (1995). Lanthanides, *in* Kroschwitz, J.I., and

Grant, M.H., eds., Kirk-Othmer encyclopedia of chemical technology, 4th ed.: *New York, John Wiley,* v. *14*, 1091–1115.

[17] Taylor, S. R. & McLennan, S. M. (1985). *The continental crust, its composition and evolution—An examination of the geochemical record preserved in sedimentary rocks*: Oxford, Blackwell Scientific, 312 p.

[18] U.S. Geological Survey. (2010). Mineral commodity summaries 2010: U.S. Geological Survey, 193 p., accessed September 8, 2010, at http://minerals.usgs.gov/minerals/pubs/mcs/.

[19] Wedepohl, K. H. (1995). The composition of the continental crust: *Geochimica et Cosmochimica Acta,* v. *59*, no. 7, 1217–1232.

[20] Wooley, A. R. & Kjarsgaard, B. A. (2008). Paragenetic types of carbonatite as indicated by the diversity and relative abundances of associated silicate rocks—Evidence from a global database: *Canadian Mineralogist*, v. *46*, no. 4, 741–752.

[21] Wu, C., Yuan, Z. & Bai, G. (1996). Rare earth deposits in China, *in* Jones, A.P., Wall, F., and Williams, C.T., eds., Rare earth minerals—Chemistry, origin, and ore deposits: London, Chapman and Hall, 281–310.

[22] Alkane Resources Ltd. (2010). The Dubbo zirconia project: Perth, Western Australia, Alkane Resources Ltd., 6 *p.* www.alkane. accessed July 15, 2010.

[23] Arafura Resources Ltd. (2010). Annual Operations Report 2009 (Year ended June 30): Perth, Western Australia, Arafura Resources Ltd., 32 p. www.arafuraresources.com.au/documents/ARU2009OperationsReportFINAL.pdf, accessed July 15, 2010

[24] Bastos Neto, A. C. & Pereira, V. P. (2009). The world- class Sn, Nb, Ta, F (Y, REE, Li) deposit and the massive cryolite associated with the albite-enriched facies of the Madeira A-type granite, Pitinga mining district, Amazonas State, Brazil: *The Canadian Mineralogist,* v. *47*, 1329–1357.

[25] Behre Dolbear. (2010). 2010 ranking of countries for mining investment: Downloaded July 13, 2010 at *www.dolbear. com/Publications/ CountryRankings.pdf.*

[26] Bogdetsky, V., Stavinskiy, V., Shukurov, E. & Suyunbaev, M. (2001). Mining industry and sustainable development in Kyrgyzstan: International Institute for Environment and Development, 95 p.

[27] Capital Mining Ltd. (2009). Annual Report 2009 (Year ended June 30): Philip, Australian Capital Territory, Capital Mining Ltd., 68 p. www.capitalmining.com.au/Annual%20Reports/

CMYj01_AnnualReport2009.pdf, accessed July 15, 2010.
[28] Chalmers, D. I. (1990). Brockman multi-metal and rare-earth deposit, *in* Hughes, F.E., ed., Geology of the mineral deposits of Australia and Papua New Guinea: *Australasian Institute of Mining and Metallurgy Monograph 14,* v. *1*, 707–709.
[29] Daigle, P. & Maunula, T. (2010). Strange Lake Project B Zone Deposit, Québec, National Instrument 43-101 resource estimate for Quest Uranium Corporation: Toronto, Ontario, Wardrop, 188 p.
[30] De Sousa, M. M. (1996). The niobium deposit in Morro dos Seis Lagos, North-Brazil: *International Geological Congress Abstracts*, v. *30*, no. 2, 784.
[31] Dunn, B. (2009). *Form 43-101 mineral resource report on the Hoidas Lake rare earth project*, Northern Mining district, Saskatchewan: Prepared for Great Western Minerals Group Ltd., Minneapolis, Minnesota, Barr, 730 p.
[32] Filho, A. I., Riffel, B. F. & de Faria Sousa, C. A. (2005). Some aspects of the mineralogy of CBMM niobium deposit and mining and pyrochlore ore processing, Araxá, MG, Brazil: Araxá, Brazil, Companhia Brasileira de Metalurgia e Mineração, 12 p.
[33] Frontier Minerals. (2009). Presentation at 5th International Rare Earth Conference, Hong Kong, November, 2009.
[34] Grauch, R. I., Verplanck, P. K., Seeger, C. M., Budahn, J. R. & Van Gosen, B. S. (2010). *Chemistry of selected core samples, concentrate, tailings, and tailings pond waters— Pea Ridge iron (-lanthanide-gold) deposit*, Washington County, Missouri: U.S. Geological Survey Open-File Report 2010–1080, 15 p.
[35] Great Western Minerals Group Ltd. (2009). *Great Western Minerals signs option agreement to recommission rare earth mine in South Africa*: News release dated January 13, 2009, accessed July 15, 2010 at www.gwmg.ca/html/media index.cfm?ReportID= 203280.
[36] Greenland Minerals and Energy Ltd. (2009). Annual Report 2009 (Year ended June 30): West Perth, Western Australia, Greenland Minerals and Energy Ltd., 112 p.
[37] Heritage Foundation. (2010). 2010 Index of Economic Freedom: Accessed July 13, 2010 at www.heritage.org/index/ *Ranking. aspx.*
[38] Hirano, H., Kamitani, M., Sato, T. & Sudo, S. (1990). Niobium mineralization of Catalão I carbonatite complex, Goias, Brazil: *Bulletin of the Geological Survey of Japan*, v. *41*, no. 11, 577–594.
[39] Hurst, C. (2010). China's rare earth elements industry—What can the

West learn?: Washington, D.C., Institute for the Analysis of Global Security, 42 p.

[40] Jackson, W. D. & Christiansen, G. (1993). International Strategic Minerals Inventory Summary Report—Rare-earth oxides: U.S. Geological Survey Circular 930–N, 68 p.

[41] Keyser, H. J. & Kennedy, J. (2007). *Geological report on the Bokan Mountain property, Prince of Wales Island, Alaska*: *Vancouver*, British Columbia, Landmark Minerals Inc., 48 p.

[42] Klemic, H., Heyl, A. V., Jr., Taylor, A. R. & Stone, J. (1959). *Radioactive rare-earth deposit at Scrub Oaks mine,* Morris County, New Jersey: U.S. Geological Survey Bulletin 1082–B, 34 p.

[43] Knox, A., Heymann, L. & Fleury, F. (2009). *Technical report on the rare earth-elements-yttrium-zirconium Zeus property*, Kipawa alkaline complex, Temiscamingue, Québec, for Matamec Explorations Inc., 159 p.

[44] Kušnir, I. (2000). Mineral resources of Vietnam: *Acta Montanistica Slovaca,* v. *5*, no. 2, 165–172.

[45] Laznicka, P. (2006). *Giant metallic deposits—Future sources of industrial minerals*: Berlin, Springer, 732 p.

[46] Long, K. R. (2009). Quantitative measures of mineral supply risk, poster presented at the American Geophysical Union, Fall Meeting, San Francisco, December, 2009.

[47] Lurie, J. (1986). Mineralization of the Pilanesberg alkaline complex, *in* Anhaeusser, C.R., and Maske, S., eds., Mineral deposits of southern Africa: *Geological Society of South Africa*, v. *2*, 2215–2228.

[48] Lynas Corporation Ltd. (2007). Lynas acquires new rare earths resource in Malawi: Press release dated September 6, 2007, 5 p. accessed July 15 at www.lynascorp.com/content/upload/ files/Announcements/Acquisition_of_New_Rare_Earths_ Resource _in _Malawi _060907 _(2) .pdf.

[49] Lynas Corporation Ltd. (2010). Will there be sufficient rare earths to meet demand from clean energy technology?: Presentation at International Minor Metals Conference, London, April, 2010, accessed July 15, 2010 at www.lynascorp.com/content/upload/files/ Presentations/MMTA_ APRIL_2010.pdf.

[50] Malyukova, N., Kim, V. & Tulyaev, R. (2005). Zonation of polymetallic, rare-earth, molybdenum, zirconium, beryllium and tantalum-niobium mineralization in the Ak-Tyuz ore deposits (northern Tien Shan), *in* Jingwen, Mao, and Bierlein, F.P., eds., Mineral deposit

research; Meeting the global challenge; *Proceedings of the 8th Biennial SGA Society for Geology Applied to Mineral Deposits Meeting: Berlin,* Springer Verlag, v. *2*, p. 1323–1326.

[51] McKeown, F. A. & Klemic, H. (1956). *Rare-earth-bearing apatite at Mineville, Essex County, New York*: U.S. Geological Survey Bulletin 1046–B, 23 p.

[52] McMahon, F. & Cervantes, M. (2010). Survey of Mining Companies 2009/2010: Toronto, Ontario, Fraser Institute, 88 p.

[53] Molycorp, Inc. (1986). Internal memorandum dated February 5, 1986, quoted by Quantum Rare Earth Developments Corp.: Accessed July 14, 2010 at www.quantumrareearth.com/elkcreek. html.

[54] Molycorp, Inc. (2010). *Prospectus dated April 16, 2010, Form S-1, Registration Statement under the Securities Act of 1933*, 108 p.

[55] Morteani, G. & Satir, M. (1989). The bastnaesite-fluoritebarite deposit of the Kizilcaören District, Eskisehhir, Turkey, *in* Möller, P., Černý, P., and Saupé, F., eds., Lanthanides, tantalum and niobium—Mineralogy, geochemistry, characteristics of primary ore deposits, prospecting, processing and applications: Berlin, Springer, p. 189–194.

[56] National Research Council. (2008). *Minerals, critical minerals, and the U.S. Economy:* Washington, D.C., National Academies Press, 264 p.

[57] Navigator Resources Ltd. (2009). 2009 Annual Report (year ending June 30): West Perth, Western Australia, Navigator Resources Ltd., 72 p.

[58] Noble, A. C., Clark, J. G. & Ranta, D. E. (2009). Technical report on the mineral resources of the Bear Lodge rare- earths project located in Crook County, Wyoming, USA, prepared for Rare Element Resources Ltd.: Lakewood, Colorado, Ore Reserves Engineering, 173 p.

[59] Northern Uranium Ltd. (2010). website, www.northernuranium. com.au/index.php/rare_earths/12, accessed July 15, 2010.

[60] Oreskes, N. & Einaudi, M. T. (1990). Origin of rare earth element-enriched hematite breccias at the Olympic Dam Cu-U-Au-Ag deposit, Roxby Downs, South Australia: *Economic Geology,* v. *85*, no. 1, p. 1–28.

[61] Organization for Economic Co-Operation and Development. (2008). Country risk classification for participants to the Arrangement on Officially Supported Export Credits, 1999 to 2009: Downloaded July 13, 2009 from www.oecd.org/ dataoecd/9/12/35483246.pdf.

[62] Orris, G. J. & Grauch, R. I. (2002). Rare earth element mines, deposits, and occurrences: U.S. Geological Survey Open- File Report 2002–0189, 174 p.

[63] Osterwald, F. W., Osterwald, D. B., Long, J. S. & Wilson, W. H. (1966). Mineral resources of Wyoming: Wyoming Geological Survey Bulletin 50, 287 p.

[64] Palmer, K. & Broad, P. (2007). *Technical report on the Thor Lake rare metals project*, NT, for Avalon Ventures Ltd.: Toronto, Ontario, Wardrop, 197 p.

[65] Papp, J. F., Bray, E. L., Edelstein, D. L., Fenton, M. D., Guberman, D. E., Hedrick, J. B., Jorgenson, J. D., Kuck, P. H., Shedd, K. B. & Tolcin, A. C. (2008). Factors that influence the price of Al, Cd, Co, Cu, Fe, Ni, Pb, rare earth elements, and Zn: U.S. Geological Survey Open-File Report 2008–1356, 61 p.

[66] Paul, J. & Stubens, T. C. (2009). Thor Lake-Lake Zone mineral resource update: Report to Avalon Rare Metals Inc., Toronto, Ontario, Wardrop, 194 p.

[67] Pell, J. (1996). Mineral deposits associated with carbonatites and related alkaline igneous rocks, *in* Mitchell, R.H., ed., Undersaturated alkaline rocks; mineralogy, petrogenesis, and economic potential: *Mineralogical Association of Canada Short Course, 24*, p. 271–310.

[68] Peters, William C. (1966). The pre-production interval of mines—Summary of an industry-wide survey: *Mining Engineering,* v. *18*, no. 8, p. 63–64.

[69] Raw Materials Supply Group. *(2010). Critical raw materials for the EU—Report of the Ad-hoc Working Group on defining critical raw materials*: Raw Materials Supply Group, European Commission, Enterprise and Industry, 220 p.

[70] Staatz, M. H., Armbrustmacher, T. J., Olson, J. C., Brownfield, I. K., Brock, M. R., Lemons, J. F., Jr., Coppa, L. V. & Clingan, B. V. (1979). *Principal thorium resources in the United States*: U.S. Geological Survey Circular 805, 42 p.

[71] Stans Energy Corporation. (2010). website, www.stansenergy. com/K _Overview.htm, accessed July 15, 2010.

[72] Stigler, George. (1983). The organization of industry: Chicago, Illinois, University of Chicago Press, 356 p.

[73] Transparency International. (2010). Corruption Perceptions Index 2009, accessed September 8, 2010, at http://www. transparency cpi_2009_table.

[74] U.S. Geological Survey. (2010). Mineral Commodity Summaries, 2010, accessed September 8, 2010, at *http://minerals.usgs.gov/minerals/ index.html.*

[75] Wedow, Helmuth, Jr. (1967). The Morro de Ferro thorium and rare-earth ore deposit, Pocos de Caldas District, Brazil: U.S. Geological Survey Bulletin 11 85–D, 34 p.

[76] World Bank. (2010). Worldwide Governance Indicators, 1996– 2008, accessed September 8, 2010, at http://info.worldbank. org/governance _chart. asp.

[77] Bell, Keith, ed. (1989). *Carbonatites—Genesis and evolution: Boston, Mass.*, Unwin Hyman Ltd., 618 p.

[78] Heinrich, E. W. (1980). *The geology of carbonatites: Huntington*, N.Y., Robert E. Krieger Publishing Company, 585 p.

[79] Sorensen, H., ed. (1974). The alkaline rocks: London, John Wiley, 622 p.

[80] Tuttle, O. F. & Gittins, J. (1966). *Carbonatites:* New York, Interscience Publishers, 591 p.

[81] Van Gosen, B. S., Gillerman, V. S. & Armbrustmacher, T. J. (2009). Thorium deposits of the United States—Energy resources for the future?: U.S. Geological Survey Circular 1336, 21 p. (Also available at http://pubs.usgs.gov/ circ/1336/.)

[82] Woolley, A. R. (1987). *Alkaline rocks and carbonatites of the World—Part I*, North and South America: Austin, Tex., University of Texas Press, 216 p.

[83] Cathrall, J. B. (1994). Geochemical survey of the Craig study area—Craig and Dixon Entrance quadrangles and the western edges of the Ketchikan and Prince Rupert quadrangles, southeast Alaska: U.S. Geological Survey Bulletin 2082, 52 p., 1 plate.

[84] De Saint-André, B., Lancelot, J. R. & Collot, B. (1983). U-Pb geochronology of the Bokan Mountain peralkaline granite, southeastern Alaska: *Canadian Journal of Earth Sciences,* v. *20*, no. 2, p. 236–245.

[85] Heylmun, E. B. (1999). Rare earths at Bokan Mountain, Alaska: *International California Mining Journal*, v. *68*, no. 5, p. 44–46.

[86] Kent, S. & Sullivan, C. (2004). 2004 Preliminary assessment/ site inspection report, Ross Adams uranium mine, Prince of Wales Island, Alaska: Private report for USDA Forest Service, Kent & Sullivan, Inc.

[87] Keyser, H. J. & McKenney, J. (2007). Geological Report on the Bokan Mountain Property, Prince of Wales Island, Alaska: Private report for Landmark Minerals, Inc., 48 p.

[88] Lanphere, M. A., MacKevett, Jr., E. M. & Stern, T. W. (1964). Potassium-argon and lead-alpha ages of plutonic rocks, Bokan Mountain area, Alaska: *Science,* v. *145*, no. 3633, p. 705–707.

[89] MacKevett, E. M., Jr. (1963). *Geology and ore deposits of the Bokan Mountain uranium-thorium area*, southeastern Alaska: U.S. Geological Survey Bulletin 1154, 125 p., 5 plates.

[90] Philpotts, J. A., Taylor, C. D., Tatsumoto, M. & Belkin, H. E. (1998). Petrogenesis of late-stage granites and Y-REE-ZrNb-enriched vein dikes of the Bokan Mountain stock, Prince of Wales Island, southeastern Alaska: U.S. Geological Survey Open-File Report 98–0459, 71 p.

[91] Staatz, M. H. (1978). I and L uranium and thorium vein system, Bokan Mountain, southeastern Alaska: *Economic Geology*, v. *73*, no. 4, p. 5 12–523.

[92] Stephens, F. H. (1971). The Kendrick Bay project: *Western Miner,* v. *44*, no. 10, p. 151–158.

[93] Thompson, T. B. (1988). Geology and uranium-thorium mineral deposits of the Bokan Mountain granite complex, southeastern Alaska: *Ore Geology Reviews*, v. *3*, p. 193–210.

[94] Warner, J. D. & Barker, J. C. (1989). *Columbium- and rare-earth element-bearing deposits at Bokan Mountain*, southeast Alaska: U.S. Bureau of Mines Open File Report 33–89, 196 p.

[95] Houston, J. R., Velikanje, R. S., Bates, R. G., & Wedow, H., Jr. (1955). *Reconnaissance for radioactive deposits in southeastern Alaska*, 1952: U.S. Geological Survey Trace Elements Investigations Report 293, 58 p.

[96] Castor, S. B. (2008). The Mountain Pass rare-earth carbonatite and associated ultrapotassic rocks, California: *The Canadian Mineralogist*, v. *46*, no. 4, p. 779–806.

[97] Castor, S. B. & Hedrick, J. B. (2006). Rare earth elements, in Kogel, J.E., Trivedi, N.C., Barker, J.M., and Krukowski, S.T., eds., Industrial minerals & rocks—Commodities, markets, and uses (7th ed.): Littleton, Colo., Society for Mining, Metallurgy, and Exploration, Inc. (SME), p. 769–792.

[98] Castor, S. B., and Nason, G. W. (2004). Mountain Pass rare earth deposit, California, in Castor, S.B., Papke, K.G., and Meeuwig, R.O., eds., Betting on industrial minerals—Proceedings of the 39th Forum on the Geology of Industrial Minerals, Reno/Sparks, Nev., May 18–24, 2003: *Nevada Bureau of Mines and Geology Special Publication 33*, p. 68–81.

[99] DeWitt, E. d., Kwak, L. M. & Zartman, R. E. (1987). U-Th-Pb and $^{40}Ar/^{39}Ar$ dating of the Mountain Pass carbonatite and alkalic igneous rocks, S.E. Cal. [abs.]: *Geological Society of America Abstracts with Programs,* v. *19*, no. 7, p. 642.

[100] Mining Engineering. (2009). Rare earths mine to come back on line to fuel green cars: *Mining Engineering*, v. *61*, no. 10, [October 2009], p. 8.

[101] Olson, J. C., Shawe, D. R., Pray, L. C. & Sharp, W. N. (1954). Rare-earth mineral deposits of the Mountain Pass district, San Bernardino County, California: U.S. Geological Survey Professional Paper 261, 75 p.

[102] Shawe, D. R. (1953). Thorium resources of the Mountain Pass district, San Bernardino County, California: U.S. Geological Survey Trace Elements Investigations Report 251, 73 p., 4 plates.

[103] Staatz, M. H., Hall, R. B., Macke, D. L., Armbrustmacher, T. J. &Brownfield, I. K. (1980). *Thorium resources of selected regions in the United States*: U.S. Geological Survey Circular 824, 32 p.

[104] Evans, J. R. (1964). Xenotime mineralization in the southern Music Valley area, Riverside County, California: California Division of Mines and Geology Special Report 79, 24 p.

[105] Armbrustmacher, T. J. (1983). The complex of alkaline rocks at Iron Hill, Powderhorn district, Gunnison County, Colorado, *in* Handfield, R.C., ed., Gunnison Gold Belt and Powderhorn Carbonatite Field Trip Guidebook, 1983: Denver Region Exploration Geologists Society, p. 28–31.

[106] Armbrustmacher, T. J. & Brownfield, I. K. (1979). The carbonatite stock at Iron Hill, Gunnison County, Colorado— Chemical and mineralogical data: U.S. Geological Survey Open-File Report 79–537, 10 p.

[107] Best, M. G. (1982). *Igneous and metamorphic petrology*: San Francisco, Calif., W.H. Freeman and Company, 630 p.

[108] Hedlund, D. C. & Olson, J. C. (1975). *Geologic map of the Powderhorn quadrangle, Gunnison and Saguache Counties*, Colorado: U.S. Geological Survey Geologic Quadrangle Map GQ–1 178, scale 1:24,000.

[109] Olson, J. C. (1974). Geologic map of the Rudolph Hill quadrangle, Gunnison, Hinsdale, and Saguache Counties, Colorado: U.S. Geological Survey Geologic Quadrangle Map GQ–1177, scale, 1, 24,000.

[110] Olson, J. C. & Hedlund, D. C. (1981). Alkalic rocks and resources of thorium and associated elements in the Powderhorn district, Gunnison County, Colorado: U.S. Geological Survey Professional Paper 1049–C, 34 p.

[111] Shaver, K. C. & Lunceford, R. A. (1998). White Earth Project, Colorado—The largest titanium resource in the United States: *Canadian Industrial Minerals Bulletin,* v. *91*, p. 63–65.

[112] Staatz, M. H., Armbrustmacher, T. J., Olson, J. C., Brownfield, I. K., Brock, M. R., Lemons, J. F., Jr., Coppa, L. V. & Clingan, B. V. (1979). Principal thorium resources in the United States: U.S. Geological Survey Circular 805, 42 p.

[113] Thompson, J. V. (1983). Economic potential of the carbonatite and alkalic stock at Powderhorn, Gunnison County, Colorado, *in* Handfield, R.C., ed., Gunnison Gold Belt and Powderhorn Carbonatite Field Trip Guidebook, 1983: Denver Region Exploration Geologists Society, p. 32–42.

[114] Thompson, J. V. (1987). Titanium resource in Colorado equals all other US deposits: *Engineering and Mining Journal*, v. *188*, no. 7, p. 27–30.

[115] Upton, G. J. (1967). Alkaline pyroxenites, *in* Wyllie, P.J., ed., Ultramafic and related rocks: New York, John Wiley, p. 28 1–288.

[116] Van Gosen, B. S. (2008). Geochemistry of rock samples collected from the Iron Hill carbonatite complex, Gunnison County, Colorado: U.S. Geological Survey Open-File Report 2008–1119, 27 p. and 2 spreadsheet data files. (Also available at *http://pubs.usgs.gov/of/2008/1119/)*

[117] Van Gosen, B. S. & Lowers, H. A. (2007). Iron Hill (Powderhorn) carbonatite complex, Gunnison County, CO—A potential source of several uncommon mineral resources: *Mining Engineering*, v. *59*, no. 10, [October 2007], p. 56–62.

[118] Armbrustmacher, T. J. (1979). Replacement and primary magmatic carbonatites from the Wet Mountains area, Fremont and Custer Counties, Colorado: *Economic Geology*, v. *74*, no. 4, p. 888–901.

[119] Armbrustmacher, T. J. (1984). *Alkaline rock complexes in the Wet Mountains area, Custer and Fremont Counties*, Colorado: U.S. Geological Survey Professional Paper 1269, 33 p.

[120] Armbrustmacher, T. J. (1988). Geology and resources of thorium and associated elements in the Wet Mountains area, Fremont and Custer Counties, Colorado: U.S. Geological Survey Professional Paper 1049–F, 34 p., 1 plate.

[121] Armbrustmacher, T. J. & Brownfield, I. K. (1978). Carbonatites in the Wet Mountains area, Custer and Fremont Counties, Colorado—Chemical and mineralogical data: U.S. Geological Survey Open-File Report 78–1 77, 6 p., 3 sheets.

[122] Armbrustmacher, T. J., Brownfield, I. K. & Osmonson, L. M. (1979). Multiple carbonatite dike at McClure Gulch, Wet Mountains alkalic province, Fremont County, Colorado: *The Mountain Geologist*, v. *16*,

no. 2, p. 37–45.

[123] Christman, R. A., Brock, M. R., Pearson, R. C. & Singewald, Q. D. (1959). Geology and thorium deposits of the Wet Mountains, Colorado—A progress report: U.S. Geological Survey Bulletin 1072–H, 535 p.

[124] Christman, R. A., Heyman, A. M., Dellwig, L. F. & Gott, G. B. (1953). Thorium investigations 1950–52, Wet Mountains, Colorado: U.S. Geological Survey Circular 290, 40 p., 5 plates.

[125] Heinrich, E. W. &Dahlem, D. H. (1967). Carbonatites and alkalic rocks of the Arkansas River area, Fremont County, Colorado, 4—The Pinon Peak breccia pipes: *American Mineralogist,* v. *52*, nos. 5 and 6, p. 817–83 1.

[126] Heinrich, E. W. & Salotti, C. A. (1975). A colloform carbonatite, McCoy Gulch, Fremont County, Colorado: *The Mountain Geologist*, v. *12*, no. 3, p. 103–111.

[127] Heinrich, E. W. & Shappirio, J. R. (1966). Alkalic rocks and carbonatites of the Arkansas River Canyon, Fremont County, Colorado, 3—The Amethyst carbonatites: *American Mineralogist,* v. *51*, no. 7, p. 1088–1106.

[128] Olson, J. C., Marvin, R. F., Parker, R. L. &Mehnert, H. H. (1977). Age and tectonic setting of lower Paleozoic alkalic and mafic rocks, carbonatites, and thorium veins in south-central Colorado: *U.S. Geological Survey Journal of Research,* v. *5*, no. 6, p. 673–687.

[129] Parker, R. L. &Sharp, W. N. (1970). *Mafic-ultramafic igneous rocks and associated carbonatites of the Gem Park Complex*, Custer and Fremont Counties, Colorado: U.S. Geological Survey Professional Paper 649, 24 p.

[130] Shawe, D. R. & Parker, R. L. (1967). *Mafic-ultramafic layered intrusion at Iron Mountain, Fremont County*, Colorado: U.S. Geological Survey Bulletin 1251–A, 28 p.

[131] Staatz, M. H. & Conklin, N. M. (1966). Rare-earth thorium carbonate veins of the Road Gulch area, northern Wet Mountains, Colorado, *in* Geological Survey research, 1966, Chapter B: U.S. Geological Survey Professional Paper 550–B, p. 130–133.

[132] Anderson, A. L. (1958). Uranium, thorium, columbium, and rare earth deposits in the Salmon region, Lemhi County, Idaho: Idaho Bureau of Mines and Geology Pamphlet No. 115, 81 p.

[133] Evans, K. V. & Green, G. N. (2003). Geologic map of the Salmon National Forest and vicinity, east-central Idaho: U.S. Geological Survey

Geologic Investigations Series Map I–2765, scale 1:100,000, 2 sheets, 19 p. pamphlet.

[134] Evans, K. V. & Zartman, R. E. (1990). U-Th-Pb and Rb-Sr geochronology of middle Proterozoic granite and augen gneiss, Salmon River Mountains, east-central Idaho: *Geological Society of America Bulletin,* v. *102*, no. 1, p. 63–73.

[135] Hedrick, J. B. (2010). Rare earths (advance release), *in* 2007 Minerals yearbook: U.S. Geological Survey, 19 p., accessed September 8, 2010, at http://minerals.usgs.gov/minerals/ pubs/commodity/rare _earths/myb1-2007-raree.pdf.

[136] Staatz, M. H., Armbrustmacher, T. J., Olson, J. C., Brownfield, I. K., Brock, M. R., Lemons, J. F., Jr., Coppa, L. V. &Clingan, B. V. (1979). *Principal thorium resources in the United States*: U.S. Geological Survey Circular 805, 42 p.

[137] Staatz, M. H. (1972). Thorium-rich veins of Hall Mountain in northernmost Idaho: *Economic Geology*, v. *67*, no. 2, p. 240–248.

[138] Staatz, M. H., Shaw, V. E. & Wahlberg, J. S. (1974). Distribution and occurrence of rare earths in the thorium veins on Hall Mountain, Idaho: *Journal of Research of the U.S. Geological Survey,* v. *2*, no. 6, p. 677–683.

[139] Staatz, M. H., Armbrustmacher, T. J., Olson, J. C., Brownfield, I. K., Brock, M. R., Lemons, J. F., Jr., Coppa, L. V. & Clingan, B. V. (1979). *Principal thorium resources in the United States*: U.S. Geological Survey Circular 805, 42 p.

[140] Hedrick, J. B. (2010). Rare earths (advance release), *in* 2007 Minerals yearbook: U.S. Geological Survey, 19 p. accessed September 8, 2010, at http://minerals.usgs.gov/minerals/ pubs/commodity/rare _earths/myb1-2007-raree.pdf.

[141] Staatz, M. H. (1972). Geology and description of the thorium- bearing veins, Lemhi Pass quadrangle, Idaho and Montana: U.S. Geological Survey Bulletin 1351, 94 p.

[142] Staatz, M. H., Armbrustmacher, T. J., Olson, J. C., Brownfield, I. K., Brock, M. R., Lemons, J. F., Jr., Coppa, L. V. & Clingan, B. V. (1979). Principal thorium resources in the United States: U.S. Geological Survey Circular 805, 42 p.

[143] Van Gosen, B. S., Gillerman, V. S. & Armbrustmacher, T. J. (2009). Thorium deposits of the United States—Energy resources for the future?: U.S. Geological Survey Circular 1336, 21 p. (Also available at http://pubs.usgs.gov/ circ/1336/.)

[144] Brown, J. S., Emery, J. A. & Meyer, P. A., Jr. (1954). Explosion pipe in test well on Hicks Dome, Hardin County, Illinois: *Economic Geology*, v. *49*, no. 8, p. 891–902.

[145] Heyl, A. V., Brock, M. R., Jolly, J. L. & Wells, C. E. (1965). *Regional structure of the southeast Missouri and Illinois- Kentucky mineral districts*: U.S. Geological Survey Bulletin 1202–B, 20 p., 3 plates.

[146] Pitkin, J. A. (1974). Preliminary thorium daughter contour map and profiles of the Hicks Dome area, Hardin County, Illinois: U.S. Geological Survey Open-File Report 74–157, 9 p.

[147] Trace, R. D. (1960). Significance of unusual mineral occurrence at Hicks Dome, Hardin County, Illinois, *in* Short papers in the geological sciences, Geological Survey research, 1960: U.S. Geological Survey Professional Paper 400–B, p. B63–B64.

[148] Castor, S. B. & Nason, G. W. (2004). Mountain Pass rare earth deposit, California, *in* Castor, S.B., Papke, K.G., and Meeuwig, R.O., eds., Betting on industrial minerals—Forum on the Geology of Industrial Minerals, 39th, Reno/Sparks, Nev., May 18–24, 2003, Proceedings: Nevada Bureau of Mines and Geology Special Publication 33, p. 68–81.

[149] Clark, A. L. & Zheng, Shuhui (1991). China's rare earth potential, industry and policy, *in* Siribumrungsukha, B., Arrykul, S., Sanguansai, P., Pungrassami, T., Sikong, L., and Kooptarnond, K., eds., International Conference on Rare Earth Minerals and Minerals for Electronic Uses, Hat Yai, Thailand, January 23–25, 1991, Proceedings: Hat Yai, Thailand, Prince of Songkla University, p. 577–601.

[150] Emery, J. A. (1968). Geology of the Pea Ridge iron ore body, *in* Ridge, J.D., ed., Ore deposits in the United States, 1933– 1967, The Graton-Sales Volume, volume 1: New York, N.Y., American Institute of Mining, Metallurgical, and Petroleum Engineers, Inc., p. 3 59–369.

[151] Grauch, R. I., Verplanck, P. L., Seeger, C. M., Budahn, J. R. & Van Gosen, B. S. (2010). Chemistry of selected core samples, concentrate, tailings, and tailings pond waters—Pea Ridge iron (-lanthanide-gold) deposit, Washington County, Missouri: U.S. Geological Survey Open-File Report 2010–1080, 15 p. (Also available at http://pubs.usgs.gov/of/2010/1080/.)

[152] Husman, J. R. (1989). Gold, rare earth element, and other potential by-products of the Pea Ridge iron ore mine: Missouri Department of Natural Resources, Division of Geology and Land Survey, Contributions to Precambrian Geology 21, Open-File Report OFR–89–78–MR, 18 p.

[153] Kisvarsanyi, E. B. (1980). Granitic ring complexes and Precambrian hot-

spot activity in the St. Francois terrane, midcontinent region, United States: *Geology*, v. *8*, no. 1, p. 43–47.

[154] Kisvarsanyi, Geza & Proctor, P. D. (1967). Trace element content of magnetites and hematites, southeast Missouri iron metallogenic province, U.S.A.: *Economic Geology*, v. *62*, no. 4, p. 449–471.

[155] Nuelle, L. M. (1998). Minerals of the Pea Ridge mine, Washington County, Missouri: *Rocks & Minerals*, v. *73*, issue 2, p. 90–97.

[156] Nuelle, L. M., Day, W. C., Sidder, G. B. & Seeger, C. M. (1992). Geology and mineral paragenesis of the Pea Ridge iron ore mine, Washington County, Missouri—Origin of the rare-earth-element- and gold-bearing breccia pipes, Chapter A, *in* Day, W.C., and Lane, D.E., eds., Strategic and critical minerals in the midcontinent region, United States: U.S. Geological Survey Bulletin 1989–A, p. A1–A11.

[157] Nuelle, L. M., Kisvarsanyi, E. B., Seeger, C. M., Day, W. C. & Sidder, G. B. (1991). Structural setting and control of the Pea Ridge magnetite deposit, Middle Proterozoic St. Francois terrane, Missouri [abs.]: *Geological Society of America Abstracts with Programs*, v. *23*, no. 5, p. 292.

[158] Seeger, C. M., Nuelle, L. M., Day, W. C., Sidder, G. B., Marikos, M. A. & Smith, D. C. (2001). Geologic maps and cross sections of mine levels at the Pea Ridge iron mine, Washington County, Missouri: U.S. Geological Survey Miscellaneous Field Studies Map MF–2353, version 1.0, 6 p. pamphlet, 5 sheets.

[159] Sidder, G. B., Day, W. C., Nuelle, L. M., Seeger, C. M. & Kisvarsanyi, E. B. (1993). Mineralogic and fluid-inclusion studies of the Pea Ridge iron-rare-earth-element deposit, southeast Missouri, Chapter U, *in* Scott, R.W., Jr., Detra, P.S., and Berger, B.R., eds., Advances related to United States and international mineral resources—Developing frameworks and exploration technologies: U.S. Geological Survey Bulletin 2039, p. 205–216.

[160] Staatz, M. H. (1972). Geology and description of the thorium- bearing veins, Lemhi Pass quadrangle, Idaho and Montana: U.S. Geological Survey Bulletin 1351, 94 p.

[161] Staatz, M. H., Armbrustmacher, T. J., Olson, J. C., Brownfield, I. K., Brock, M. R., Lemons, J. F., Jr., Coppa, L. V. & Clingan, B. V. (1979). *Principal thorium resources in the United States*: U.S. Geological Survey Circular 805, 42 p.

[162] Van Gosen, B. S. (2008). Geochemistry of rock samples collected from the Iron Hill carbonatite complex, Gunnison County, Colorado: U.S.

Geological Survey Open-File Report 2008–1119, 27 p. and 2 spreadsheet data files. (Also available at *http://pubs.usgs.gov/of/2008/1119/*)

[163] Vierrether, C. W. & Cornell, W. L. (1993). Rare-earth occurrences in the Pea Ridge tailings: U.S. Bureau of Mines Report of Investigations 9453, 10 p.

[164] Whitten, C. W. & Yancey, R. J. (1990). Characterization of the rare-earth mineralogy at the Pea Ridge deposit, Missouri: U.S. Bureau of Mines Report of Investigations 9331, 9 p.

[165] Yang, Xiao-Yong, Sun, Wei-Dong, Zhang, Yu-Xu & Zheng, Yong-Fei. (2009). Geochemical constraints on the genesis of the Bayan Obo Fe-Nb-REE deposit in Inner Mongolia, China: *Geochimica et Cosmochimica Acta*, v. *73*, p. 1417–1435.

[166] Burfeind, W. J., Carlson, M. P. & Smith, Russell. (1971). The Elk Creek geophysical anomaly, Johnson and Pawnee Counties, Nebraska [abs.]: *Geological Society of America Abstracts with Programs*, v. *3*, no. 4, p. 254.

[167] Carlson, M. P. & Treves, S. B. (2005). The Elk Creek Carbonatite, southeast Nebraska—An overview: *Natural Resources Research,* v. *14*, no. 1, p. 39–45.

[168] Xu, Anshun. (1996). Mineralogy, petrology, geochemistry and origin of the Elk Creek Carbonatite, Nebraska: Lincoln, Nebr., University of Nebraska, Ph.D. dissertation, 299 p.

[169] Griswold, G. B. (1959). Mineral deposits of Lincoln County, New Mexico: *New Mexico Bureau of Mines and Mineral Resources Bulletin, 67*, 117 p., 12 plates.

[170] McLemore, V. T. (1983). Uranium and thorium occurrences in New Mexico—Distribution, geology, production, and resources, with selected bibliography: New Mexico Bureau of Mines and Mineral Resources Open-File Report 183, 180 p. text, 6 appendixes.

[171] McLemore, V. T., North, R. M. & Leppert, Shawn. (1988). REE, niobium, and thorium districts and occurrences in New Mexico: New Mexico Bureau of Mines and Mineral Resources Open-File Report 324, 27 p., 2 plates.

[172] Staatz, M. H. (1974). Thorium veins in the United States: *Economic Geology,* v. *69*, no. 4, p. 494–507.

[173] McLemore, V. T., North, R. M. & Leppert, Shawn. (1988). REE, niobium, and thorium districts and occurrences in New Mexico: *New Mexico Bureau of Mines and Mineral Resources Open-File Report, 324*,

27 p., 2 plates.

[174] Robertson, J. M. (1976). Mining districts of northeastern New Mexico, *in* Ewing, R.C., and Kues, B.S., eds., Guidebook of Vermejo Park, northeastern New Mexico, New Mexico Geological Society Twenty-seventh Field Conference, September 30, October 1 and 2, 1976: New Mexico Geological Society, p. 257–262.

[175] DeMark, R. S. (1980). The Red Cloud mines, Gallinas Mountains, New Mexico: *The Mineralogical Record,* v. *2*, no. 11, p. 69–72.

[176] Glass, J. J. & Smalley, R. G. (1945). Bastnasite: *American Mineralogist,* v. *30*, nos. 9 and 10, p. 601–615.

[177] Griswold, G. B. (1959). Mineral deposits of Lincoln County, New Mexico: *New Mexico Bureau of Mines and Mineral Resources Bulletin, 67*, 117 p., 12 plates.

[178] Perhac, R. M. (1970). Geology and mineral deposits of the Gallinas Mountains, Lincoln and Torrance Counties, New Mexico: *New Mexico Bureau of Mines and Mineral Resources Bulletin, 95*, 51 p., 2 plates.

[179] Perhac, R. M. & Heinrich, E. W. (1964). Fluorite-bastnaesite deposits of the Gallinas Mountains, New Mexico and bastnaesite paragenesis: *Economic Geology*, v. *59*, no. 2, p. 226–239.

[180] Soule, J. H. (1946). Exploration of Gallinas fluorspar deposits, Lincoln County, N. Mex.: U.S. Bureau of Mines Report of Investigations 3854, 25 p.

[181] Gillerman, Elliott. (1964). Mineral deposits of western Grant County, New Mexico: New Mexico Bureau of Mines and Mineral Resources Bulletin 83, 213 p., 11 plates.

[182] Hedlund, D. C. (1978). Geologic map of the Gold Hill quadrangle, Hidalgo and Grant Counties, New Mexico: U.S. Geological Survey Miscellaneous Field Studies Map MF–1035, scale 1:24,000.

[183] Richter, D. H. & Lawrence, V. A. (1983). *Mineral deposit map of the Silver City 1° x 2° quadrangle*, New Mexico and Arizona: U.S. Geological Survey Miscellaneous Investigations Series Map I–1310–B, scale 1:250,000.

[184] Richter, D. H., Sharp, W. N., Watts, K. C., Raines, G. L., Houser, B. B. & Klein, D. P. (1986). *Maps showing mineral resource assessment of the Silver City 1° x 2° quadrangle*, New Mexico and Arizona: U.S. Geological Survey Miscellaneous Investigations Series Map I–131 0–F, 24 p. pamphlet, 1 sheet, various scales.

[185] Staatz, M. H. (1974). Thorium veins in the United States: *Economic Geology*, v. *69*, no. 4, p. 494–507.

[186] Staatz, M. H. (1985). Geology and description of the thorium and rare-earth veins in the Laughlin Peak area, Colfax County, New Mexico: U.S. Geological Survey Professional Paper 1049–E, 32 p., 1 plate, scale 1:12,000.
[187] Staatz, M. H. (1986). Geologic map of the Pine Buttes quadrangle, Colfax County, New Mexico: U.S. Geological Survey Geologic Quadrangle Map GQ–1 591, scale 1:24,000.
[188] Staatz, M. H. (1987). Geologic map of the Tres Hermanos Peak quadrangle, Colfax County, New Mexico: U.S. Geological Survey Quadrangle Map GQ–1 605, scale 1:24,000.
[189] McLemore, V. T. (1983). Carbonatites in the Lemitar and Chupadera Mountains, Socorro County, New Mexico, *in* Chapin, C.E., ed., Socorro Region II, New Mexico Geological Society Thirty-fourth Annual Field Conference, October 13–15, 1983: New Mexico Geological Society, p. 23 5–240.
[190] McLemore, V. T. (1987). Geology and regional implications of carbonatites in the Lemitar Mountains, central New Mexico: *Journal of Geology*, v. *95*, no. 2, p. 255–270.
[191] McLemore, V. T., North, R. M. & Leppert, Shawn. (1988). REE, niobium, and thorium districts and occurrences in New Mexico: *New Mexico Bureau of Mines and Mineral Resources Open-File Report, 324*, 27 p., 2 plates.
[192] Van Allen, B. R., Emmons, D. L. & Paster, T. P. (1986). Carbonatite dikes of the Chupadera Mountains, Socorro County, New Mexico: *New Mexico Geology,* v. *8*, no. 2, p. 25–29.
[193] Bingler, E. C. (1968). Geology and mineral resources of Rio Arriba County, New Mexico: New Mexico Bureau of Mines and Mineral Resources Bulletin 91, 158 p., 8 plates.
[194] Jahns, R. H. (1946). Mica deposits of the Petaca district, Rio Arriba County, New Mexico: *New Mexico Bureau of Mines and Mineral Resources Bulletin, 25*, 294 p., 25 plates.
[195] McLemore, V. T., North, R. M. & Leppert, Shawn (1988). REE, niobium, and thorium districts and occurrences in New Mexico: New Mexico Bureau of Mines and Mineral Resources Open-File Report 324, 27 p., 2 plates.
[196] Redmon, D. E. (1961). *Reconnaissance of selected pegmatite districts in north-central New Mexico*: U.S. Bureau of Mines Information Circular 8013, 79 p.
[197] Wright, L. A. (1948). The Globe pegmatite, Rio Arriba County, New

Mexico: *American Journal of Science*, v. *246*, no. 11, p. 665–688.
[198] McLemore, V. T. (1983). Uranium and thorium occurrences in New Mexico—Distribution, geology, production, and resources, with selected bibliography: *New Mexico Bureau of Mines and Mineral Resources Open-File Report, 183*, 180 p. text, 6 appendixes.
[199] McLemore, V. T. (1986). Geology, geochemistry, and mineralization of syenites in the Red Hills, southern Caballo Mountains, Sierra County, New Mexico—Preliminary observations, *in* Clemons, R.E., King, W.E., and Mack, G.H., eds., Truth or Consequences region, New Mexico Geological Society Thirty-seventh Annual Field Conference, October 16–18, 1986: New Mexico Geological Society, p. 151–1 59.
[200] Staatz, M. H., Adams, J. W. & Conklin, N. M. (1965). Thorium-bearing microcline-rich rocks in the southern Caballo Mountains, Sierra County, New Mexico, *in* Geological Survey Research 1965, Chapter D: U.S. Geological Survey Professional Paper 525–D, p. 48–51.
[201] Holser, W. T. (1959). Trans-Pecos region, Texas and New Mexico, *in* Warner, L.A., Holser, W.T., Wilmarth, V.R., and Cameron, E.N., Occurrence of nonpegmatite beryllium in the United States: U.S. Geological Survey Professional Paper 318, p. 130–143.
[202] McLemore, V. T. (1983). Uranium and thorium occurrences in New Mexico—Distribution, geology, production, and resources, with selected bibliography: New Mexico Bureau of Mines and Mineral Resources Open-File Report 183, 180 p. text, 6 appendixes.
[203] McLemore, V. T., North, R. M. & Leppert, Shawn. (1988). REE, niobium, and thorium districts and occurrences in New Mexico: *New Mexico Bureau of Mines and Mineral Resources Open-File Report, 324*, 27 p., 2 plates.
[204] Blake, W. P. (1858). Lanthanite and allanite in Essex County, N.Y.: *American Journal of Science and Arts, Series 2,* v. *26*, no. 76, p. 245–246.
[205] Kemp, J. F. (1908). The Mineville-Port Henry group, *in* Newland, D.H., and Kemp, J.F., Geology of the Adirondack magnetic iron ores: New York State Museum Bulletin 119, p. 57–88.
[206] Lupulescu, Marion & Pyle, Joseph. (2008). Mining history, mineralogy and origin of the gneiss (granite)-hosted Fe-PREE and Fe oxide and gabbro-hosted Ti-Fe oxide deposits from the Mineville-Port Henry Region, Essex County, NY, *in* Selleck, B.W., ed., Field trip guidebook for the 80th annual meeting of the New York State Geological Association: New York, N.Y., Guidebook-New York State Geological

Association, v. 80, p. 1 17–129.

[207] McKeown, F. A. & Klemic, Harry. (1956). Rare-earth-bearing apatite at Mineville, Essex County, New York: U.S. Geological Survey Bulletin 1046–B, p. 9–23.

[208] Staatz, M. H., Hall, R. B., Macke, D. L., Armbrustmacher, T. J. & Brownfield, I. K. (1980). *Thorium resources of selected regions in the United States*: U.S. Geological Survey Circular 824, 32 p.

[209] Ranta, Don & Clark, Jim. (2010). Geology and evaluation of the Bear Lodge rare-earths project, Wyoming [unpublished abs.]: Presentation to the Denver Region Exploration Geologists' Society, January 4, 2010, Lakewood, Colo.

[210] Staatz, M. H. (1983). Geology and description of thorium and rare-earth deposits in the southern Bear Lodge Mountains, northeastern Wyoming: U.S. Geological Survey Professional Paper 1049–D, 52 p., 2 plates.

[211] Staatz, M. H., Armbrustmacher, T. J., Olson, J. C., Brownfield, I. K., Brock, M. R., Lemons, J. F., Jr., Coppa, L. V. & Clingan, B. V. (1979). *Principal thorium resources in the United States*: U.S. Geological Survey Circular 805, 42 p.

[212] Altschuler, Z. S., Cisney, E. A. & Barlow, I. H. (1952). X-ray evidence of the nature of carbonate-apatite: *Bulletin of the Geological Society of America*, v. *63*, p. 1230–1231.

[213] Cathcart, J. B. (1949). *Distribution of uranium in the Florida phosphate field*: U.S. Geological Survey Trace Elements Investigations Report 85, 18 p.

[214] Cathcart, J. B., Blade, L. V., Davidson, D. F. & Ketner, K. B. (1952). The geology of the Florida land-pebble phosphate deposits: U.S. Geological Survey Trace Elements Investigations Report 265, 21 p.

[215] Clarke, R. S., Jr. & Altschuler, Z. S. (1958). Determination of the oxidation state of uranium in apatite and phosphorite deposits: *Geochimica et Cosmochimica Acta*, v. *13*, p. 127–142.

[216] Florida Institute of Phosphate Research (FIPR). (2010). accessed September 8, 2010, at http://www.fipr.state.fl.us/ research-area-mining.htm#History.

[217] Jasinski, S. M. (2010). Phosphate rock: U.S. Geological Survey Mineral Commodity Summaries 2010, p. 118–119, accessed October 6, 2010, at http://minerals.usgs.gov/minerals/pubs/ commodity/phosphate_rock/mcs-2010-phosp .pdf.

[218] McArthur, J. M. & Walsh, J. N. (1984). Rare-earth geochemistry of phosphorites: *Chemical Geology*, v. *47*, p. 191–220.

[219] McKelvey, V. E., Cathcart, J. B. & Worthing, H. W. (1951). Preliminary note on the minor-metal content of Florida phosphate rock: U.S. Geological Survey Trace Elements Memorandum Report 236, 6 p.

[220] Riggs, S. R. (1980). Intraclast and pellet phosphorite sedimentation in the Miocene of Florida: *Journal of the Geological Society*, v. *137*, p. 741–748.

[221] Riggs, S. R. (1984). Paleoceanographic model of Neogene phosphorite deposition, *U.S. Atlantic Continental Margin: Science*, v. *223*, no. 4632, p. 123–13 1.

[222] Stowasser, W. F. (1975). Phosphate rock, *in* Minerals Yearbook 1973, v. 1, Metals, Minerals, and Fuels: U.S. Bureau of Mines, p. 1019–1035.

[223] Van Kauwenbergh, S. J. & McClellan, G. H. (1990). Comparative geology and mineralogy of the southern United States and Togo phosphorites: *Geological Society of London, Special Publications*, v. *52*, p. 139–1 55.

[224] Zanin, Y. N. & Zamirailova, A. G. (2009). Rare earth elements in supergene phosphorites: *Geochemistry International,* v. *47*, p. 282–296.

[225] Bhola, K. L., Chatterji, B. D., Dar, K. K., Mahadevan, C., Mahedevan, V., Mehta, N. R., Handi, N., Nanhi, H., Narayandas, G. R., Sahasrabudhe, G. H., Shirke, V. G. & Udas, G. R. (1958). A survey of uranium and thorium occurrences in India, *in Proceedings of the Second United Nations International Conference on the Peaceful Uses of Atomic Energy*, Geneva, Switzerland, September 1–13, 1958—Volume 2, Survey of raw material resources: Geneva, United Nations Publication, p. 100–102.

[226] Mahadevan, V., Narayana Das, G. R. & Nagaraja Roa, N. (1958). Prospecting and evaluation of beach placers along the Coastal Belt of India, *in Proceedings of the Second United Nations International Conference on the Peaceful Uses of Atomic Energy*, Geneva, Switzerland, September 1–13, 1958—Volume *2*, Survey of raw material resources: Geneva, United Nations Publication, p. 103–106.

[227] Staatz, M. H., Armbrustmacher, T. J., Olson, J. C., Brownfield, I. K., Brock, M. R., Lemons, J. F., Jr., Coppa, L. V. & Clingan, B. V. (1979). *Principal thorium resources in the United States*: U.S. Geological Survey Circular 805, 42 p.

[228] Staatz, M. H., Hall, R. B., Macke, D. L., Armbrustmacher, T. J. & Brownfield, I. K. (1980). *Thorium resources of selected regions in the United States*: U.S. Geological Survey Circular 824, 32 p.

[229] Argall, G. O., Jr. (1954). New dredging techniques recover Idaho

monazite: *Mining World*, v. *16*, no. 2, p. 26–30.

[230] Lindgren, W. (1897). Monazite from Idaho: *The American Journal of Science, Fourth Series,* v. *4*, p. 63–64.

[231] Mackin, J. H. & Schmidt, D. L. (1957). Uranium and thorium-bearing minerals in placer deposits in Idaho: Idaho Bureau of Mines and Geology, Mineral Resources Report 7, 9 p.

[232] Savage, C. N. (1961). *Economic geology of central Idaho black- sand placers*: Idaho Bureau of Mines and Geology Bulletin 17, 160 p.

[233] Staatz, M. H., Hall, R. B., Macke, D. L., Armbrustmacher, T. J. & Brownfield, I. K. (1980). *Thorium resources of selected regions in the United States*: U.S. Geological Survey Circular 824, 32 p.

[234] Storch, R. H. & Holt, D. C. (1963). Titanium placer deposits of Idaho: U.S. Bureau of Mines Report of Investigations No. 6319, 69 p.

[235] Grosz, A. E. (1993). Use of geochemical surveys in Ti-Hf-REETh-U placer exploration—A mid-Atlantic-States example (Chapter R), *in* Scott, R.W., Jr., Detra, P.S., and Berger, B.R., eds., Advances related to United States and international mineral resources—Developing frameworks and exploration technologies: U.S. Geological Survey Bulletin 2039, p. 181–188.

[236] Kline, M. H., Griffith, R. F. & Hansen, L. A. (1954). Hollow Creek monazite placer, Aiken County, South Carolina: U.S. Atomic Energy Commission Report RME–3 127, 29 p.

[237] Mertie, J. B., Jr. (1975). *Monazite placers in the southeastern Atlantic States*: U.S. Geological Survey Bulletin 1390, 41 p.

[238] Olson, J. C. & Overstreet, W. C. (1964). *Geologic distribution and resources of thorium*: U.S. Geological Survey Bulletin 1204, 61 p.

[239] Overstreet, W. C. (1967). *The geologic occurrence of monazite*: U.S. Geological Survey Professional Paper 530, 327 p., 2 plates.

[240] Overstreet, W. C., Theobald, P. K., Jr. & Whitlow, J. W. (1959). Thorium and uranium resources in monazite placers of the western Piedmont, North and South Carolina: *Mining Engineering,* v. *11*, no. 7, p. 709–714.

[241] Overstreet, W. C., White, A. M., Whitlow, J. W., Theobald, P. K., Jr., Caldwell, D. W. & Cuppels, N. P. (1968). Fluvial monazite deposits in the southeastern United States: U.S. Geological Survey Professional Paper 568, 85 p., 9 plates.

[242] Staatz, M. H., Armbrustmacher, T. J., Olson, J. C., Brownfield, I. K., Brock, M. R., Lemons, J. F., Jr., Coppa, L. V. & Clingan, B. V. (1979). *Principal thorium resources in the United States*: U.S. Geological

Survey Circular 805, 42 p.
[243] Williams, Lloyd. (1967). Heavy minerals in South Carolina: Columbia, S.C., *South Carolina State Development Board, Division of Geology*, Bulletin No. 35, 35 p.
[244] Calver, J. L. (1957). *Mining and mineral resources: Florida Geological Survey Bulletin, 39*, 132 p.
[245] Doering, J. A. (1960). Quaternary surface formations of southern part of Atlantic Coastal plain: *Journal of Geology,* v. *68*, no. 2, p. 182–202.
[246] Force, E. R. (1991). Geology of titanium-mineral deposits: Geological Society of America Special Paper 259, 112 p.
[247] Force, E. R., Grosz, A. E., Loferski, P. J. & Maybin, A. H. (1982). Aeroradioactivity maps in heavy-mineral exploration—Charleston, South Carolina, area: U.S. Geological Survey Professional Paper 1218, 19 p., 2 plates.
[248] Grosz, A. E. (1983). Application of total-count aeroradiometric maps to the exploration for heavy-mineral deposits in the Coastal Plain of Virginia: U.S. Geological Survey Professional Paper 1263, 20 p., 5 plates.
[249] Grosz, A. E., Cathcart, J. B., Macke, D. L., Knapp, M. S., Schmidt, Walter & Scott, T. M. (1989). Geologic interpretation of the gamma-ray aeroradiometric maps of central and northern Florida: U.S. Geological Survey Professional Paper 1461, 48 p., 5 plates.
[250] Kremers, H. E. (1958). Commercial thorium ores: Society of Mining Engineers of American Institute of Mining Engineers preprint 589A18, p. 1–14.
[251] MacNeil, F. S. (1950). Pleistocene shore lines in Florida and Georgia: U.S. Geological Survey Professional Paper 221–F, p. 95–1 07, 7 plates.
[252] Mertie, J. B., Jr. (1953). Monazite deposits of the southeastern Atlantic States: U.S. Geological Survey Circular 237, 31 p., 1 plate.
[253] Mertie, J. B., Jr. (1975). Monazite placers in the southeastern Atlantic States: U.S. Geological Survey Bulletin 1390, 41 p.
[254] Neiheisel, James. (1962). Heavy-mineral investigation of Recent and Pleistocene sands of Lower Coastal Plain of Georgia: *Geological Society of America Bulletin*, v. *73*, no. 3, p. 365–374.
[255] Overstreet, W. C. (1967). *The geologic occurrence of monazite*: U.S. Geological Survey Professional Paper 530, 327 p., 2 plates.
[256] Owens, J. P., Grosz, A. E. & Fisher, J. C. (1989). *Aeroradiometric map and geologic interpretation of part of the Florence and Georgetown 1° x 2° quadrangles, South Carolina*: U.S. Geological Survey Miscellaneous

Investigations Series Map I–1948–B, 1 sheet, scale 1:250,000.

[257] Staatz, M. H., Hall, R. B., Macke, D. L., Armbrustmacher, T. J. & Brownfield, I. K. (1980). *Thorium resources of selected regions in the United States:* U.S. Geological Survey Circular 824, 32 p.

[258] Winkler, C. D. & Howard, J. D. (1977). Correlation of tectonically deformed shorelines on the southern Atlantic coastal plain: *Geology*, v. 5, no. 2, p. 123–127.

In: Rare Earth Elements in the U.S. and China ISBN: 978-1-61324-500-2
Editors: Joseph A. Meyer et al.

Chapter 2

CHINA'S RARE-EARTH INDUSTRY

Pui-Kwan Tse

INTRODUCTION

China's dominant position as the producer of over 95 percent of the world output of rare-earth minerals and rapid increases in the consumption of rare earths owing to the emergence of new clean- energy and defense-related technologies, combined with China's decisions to restrict exports of rare earths, have resulted in heightened concerns about the future availability of rare earths. As a result, industrial countries such as Japan, the United States, and countries of the European Union face tighter supplies and higher prices for rare earths. This paper briefly reviews China's rare-earth production, consumption, and reserves and the important policies and regulations regarding the production and trade of rare earths, including recently announced export quotas.

The 15 lanthanide elements—lanthanum, cerium, praseodymium, neodymium, promethium, samarium, europium, gadolinium, terbium, dysprosium, holmium, erbium, thulium, ytterbium, and lutetium (atomic numbers 57–71)—were originally known as the rare earths from their occurrence in oxides mixtures. Recently, some researchers have included two other elements—scandium and yttrium—in their discussion of rare earths. Yttrium (atomic number 39), which lies above lanthanum in transition group III of the periodic table and has a similar 3+ ion with a noble gas core, has both atomic and ionic radii similar in size to those of terbium and dysprosium and is generally found in nature with lanthanides. Scandium (atomic number 21) has a smaller ionic radius than yttrium and the lanthanides, and its chemical behavior is intermediate between that of aluminum and the lanthanides. It is found in nature with the lanthanides and yttrium.

Rare earths are used widely in high-technology and clean-energy products because they impart special properties of magnetism, luminescence, and strength. Rare earths are also used in weapon systems to obtain the same properties.

CHINA'S RESOURCES OF RARE EARTHS

China is rich in rare-earth resources and produces a number of different kinds of rare-earth products. China's scientists discovered rare-earth resources in Bayan Obo in 1927 and started the production of rare-earth concentrates in 1957. After more than 8 decades of exploration, rare-earth resources have been discovered in 21 of China's Provinces and Autonomous Regions—Fujian, Gansu, Guangdong, Guangxi, Guizhou, Hainan, Henan, Hubei, Hunan, Jiangxi, Jilin, Liaoning, Nei Mongol, Qinghai, Shaanxi, Shandong, Shanxi, Sichuan, Xinjiang, Yunnan, and Zhejiang.

As of 1998, the country's total rare-earth resource (资源量) was reported to be 92 million metric tons (Mt) (Wen, 1998, p. 140). In 2009, China's Ministry of Land and Resources (MLR) reported that the country had rare-earth reserves (储量) of 18.6 Mt in rare-earth-oxide (REO) content (China National Bureau of Statistics, 2010, p. 410).

China's reserve and resource system divides the country's minerals into reserves (储量) and resources (资L量) (China National Material and Technical Control Bureau, 1999). Reserves are divided into proved extractable reserves (可采储量) (code 111), probable extractable reserves (预可采储量) (code 121), and basic reserves (基础储量) (code 122). The discovered mineral resources are divided into measured, indicated, and inferred. Comparisons with different reserve systems are difficult and always approximate. The basic reserve (基础储量) in the Chinese reserve classification system best approximates reserves as defined by the U.S. Bureau of Mines and U.S. Geological Survey (1980) in USGS Circular 831, "Principles of a Resource/Reserve Classification for Minerals." The MLR publishes the country's reserve and resource data yearly, but these reports are not available to overseas researchers. Estimation of China's rare-earth basic reserve (基础储量) was 55 Mt in REO (China National Material and Technical Control Bureau, 1999).

CHINA'S PRODUCTION OF RARE EARTHS

China's main rare-earth production takes place in the Provinces of Fujian, Guangdong, Jiangxi, and Sichuan and in Nei Mongol Autonomous Region. Nei Mongol accounted for between 50 percent and 60 percent of China's total rare-earth concentrate output during the past decade. Sichuan Province was the second leading rare-earth concentrate producer, accounting for between 24 percent and 30 percent of production during the past decade. The remaining output was from the Provinces of Fujian, Guangdong, and Jiangxi, which are important for their production of "heavy" rare earths (Chen, 2010; Geological Publishing House, 2010).

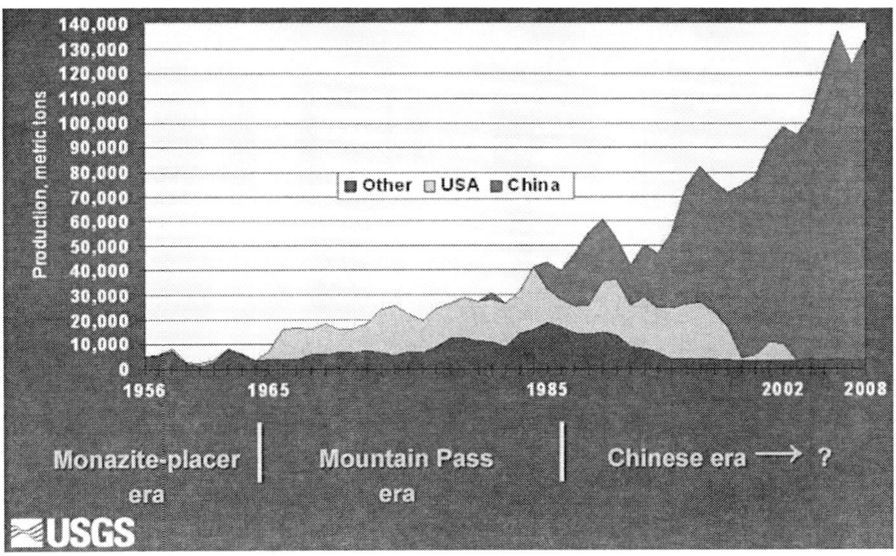

Figure 1. Global rare-earth-oxide production trends. The Mountain Pass deposit is in California, U.S.A. Graph from D.J. Cordier (U.S. Geological Survey, written commun., 2011) was updated from Haxel and others (2002, figure 1).

Table 1. China's Rare-Earth Production, Consumption, and Export Quotas for 2000 through 2011

[In metric tons of rare-earth-oxide (REO) equivalent except as described in footnote 1. Abbreviations used: est., estimated; MIIT; China Ministry of Industry and Information Technology; MLR, China Ministry of Land and Resources; NA, not available. Sources: China Ministry of Commerce (2006, 2008a,b, 2009, 201 0a,b,c)]

		2000	2001	2002	2003	2004	2005	2006	2007	2008	2009	2010	2011
Production quota	MIIT									119,500	110,700	89,200	NA
	MLR	55,000	NA	NA	NA	NA	NA	86,620	87,020	90,180	87,620	89,200	NA
Production	(est.)	73,000	81,000	88,000	92,000	98,000	119,000	133,000	120,000	125,000	129,000	120,000	NA
Consumption	(est.)	19,000	20,000	22,000	30,000	34,000	52,000	63,000	73,000	67,700	73,000	77,000	NA
Export quota	Domestic producers and traders[1]	47,000	45,000	NA	40,000	45,000	48,010	45,000	43,574	34,156	31,310	22,512[2]	14,446[3]
	Sino-foreign joint ventures[1]	NA	NA	NA	NA	NA	17,570	16,070	16,069	15,834	16,845	7,746[4]	

[1]Export quotas are in gross weight rather than REO content for domestic producers and traders and Sino-foreign producers in 2005 and later years.

[2]Total export quota for 22 domestic producers and traders.

[3]Export quota for 2011 is only for the first tranche for 31 domestic producers and traders and Sino-foreign joint-venture producers.

[4]Total export quota for Sino-foreign producers.

Between 1990 and 2000, China's production increased over 450 percent to 73,000 metric tons (t) from about 16,000 t (figure 1). During the same period, production from other countries declined almost 60 percent to about 16,000 t from 44,000 t. As a result, world production increased just over 150 percent to almost 91,000 t from about 60,000 t between 1990 and 2000. Since 2000, world and Chinese production have continued to increase; by 2009, world production increased 45 percent to about 132,000 t, and Chinese production increased 77 percent to 129,000 t. Production from other countries decreased to about 3,000 t in 2009.

The volume of China's rare-earth output as a percentage of total world output increased to more than 90 percent in 2008 from 27 percent in 1990. During the past 15 years, China has supplied more than 80 percent of the world's rare earths as concentrates, intermediate products, and chemicals. Although Chinese production expanded rapidly, China's rare-earth producers reportedly have struggled to maintain profitability. Throughout the 1 990s and 2000s, the Government and rare-earth producers had met and discussed ways to control production and exports as a means of conserving the country's mineral resources and protecting the environment. Nevertheless, competition among local governments and enterprises resulted in sustained high levels of production. Local governments depended upon rare-earth producers to provide employment and revenue for local economic development and did not always follow the national Government's guidelines on rare earths. As a result, the country's actual output of rare-earth concentrate continued to exceed the Government's production target (China State Council, 2006). Since 2006, the Government has stepped up enforcement of its policies and regulations and shut down illegal mines in the Provinces of Guangdong, Jiangxi, and Sichuan. Consequently, production of rare earths has been approximately level during the past 5 years (table 1).

China's Consumption of Rare Earths

Although China's production of rare-earth minerals increased markedly between 1990 and 2000, China's consumption of rare earths did not undergo a similar increase until after 2000. In 1990, Chinese consumption of rare earths was relatively small; however, by 2000, China's consumption had reached about 19,000 t, and world consumption was about 91,000 t. If production and consumption were about equal, then consumption was about 72,000 t by the rest of the world. From 2000 to 2009, China's consumption of rare earths increased rapidly; by 2005, China's consumption had reached almost 52,000 t and, by 2009, China's rare earth consumption was 73,000 t, about 380 percent higher than in 2000. The 2009 production and consumption data suggest that consumption of rare earths by the rest of the world has decreased to about 59,000 t. Given the rapid development of new products using rare earths, this seems unlikely and suggests that some production, possibly 20,000 t to 40,000 t, has not been accounted for in publicly available production statistics.

In China, the magnet sector accounted for about 30 percent of the total rare-earth use followed by metallurgical, 15 percent; chemical and petroleum, 10 percent; ceramics and glass, 10 percent; agriculture and textile, 10 percent; hydrogen storage, 9 percent; and others, 16 percent. In 1987, rare- earth applications in the new material industry, which includes the production of such products as catalysts, magnets, phosphors, and polishing powder, accounted for about 1 percent of the total mineral consumed; by 2008, however, it increased to more than 53 percent. In China, rare-earth consumption by the new material sector is expected to increase faster than the other traditional industrial sectors.

CHINA'S RARE-EARTH POLICIES

In 1990, the Chinese Government declared rare earths to be a protected and strategic mineral. As a consequence, foreign investors are prohibited from mining rare earths and are restricted from participating in rare-earth smelting and separation projects except in joint ventures with Chinese firms. All projects for rare-earth mining and smelting, whatever their size, required approval from the State Development and Planning Commission (SDPC) (formerly the State Planning Commission), which included the Rare-Earth Office. Sino-foreign joint-venture projects had to be approved by both SDPC and the Ministry of Commerce (MOC); prior to 2003, the Ministry of Foreign Trade and Economic Cooperation (MOFTEC) approved Sino-foreign joint ventures in rare earths.

Since the early 1990s, MLR has been responsible for developing production plans for the country's strategic commodities, including rare earths. The plans include overall production quotas, as well as quotas for individual Provinces. Provincial governments are responsible for managing their Province's allocated quota and for assigning output quotas to individual mining companies. Despite production quotas, the output of rare earths and other commodities has been much higher than the Government's target. In addition, a significant amount of the over-quota production was by miners who worked without proper mining licenses and who used obsolete mining technology that caused significant environmental damage, especially in the Provinces of Guangdong, Jiangxi, and Sichuan.

In 2008, the Rare-Earth Office was transferred to the Ministry of Industry and Information Technology (MIIT), which started issuing a rare-earth production quota for the country. The MIIT quota was much higher than the one set by MLR (table 1); this caused confusion among rare-earth producers. During 2008 and 2009, there was an apparent lack of communication between the MIIT and MLR about which agency should assign the country's rare-earth production quota. The MIIT rare-earth production quota appeared to be more reflective of rare-earth output of the country (table 1). This confusion came to an end in 2010 when both agencies set the same rare-earth production quota.

CHINA'S RARE-EARTH TRADE

In addition to setting production quotas for rare earths, China also sets quotas on the amount of rare earths that can be exported. These quotas are released in two tranches. Before 2003, rare-earth export quotas were determined by the SDPC and were distributed by the State Economic and Trade Commission (SETC), and the MOFTEC issued the licenses. After the SETC and MOFTEC were abolished, the MOC took over responsibility for the issuance and distribution of export licenses. Separate export quotas are set for domestic rare-earth producers and for Sino-foreign joint-venture rare- earth producers. Sino-foreign joint-venture rare-earth companies are allowed to export their own products under a licensing system.

Owing to the increase in domestic demand, the Government has gradually reduced the export quota during the past several years. In 2006, the Government allowed 47 domestic rare-earth producers and traders and 12 Sino-foreign rare-earth producers to export rare-earth products. In 2009, only 23 domestic rare-earth producers and traders and 11 Sino-foreign rare-earth producers were authorized to export rare-earth products. The number of permitted exporters was reduced to 22 domestic rare-earth producers and traders and 10 Sino-foreign rare-earth producers in 2010 and to 22 domestic rare-earth producers and traders and 9 Sino-foreign rare-earth producers in 2011.

In addition to setting export quotas, the Government has encouraged the export of high-value downstream products and discouraged the export of raw material. The desire to encourage exports of finished products is reflected by a change in the way the export quota was released. Initially the export quota was released as rare-earth-oxide equivalent. During the past 6 years, the Government has specified the tonnage limit of specific rare-earth products that Sino-foreign joint ventures are allowed to export, such as oxides, salts, or metal. However, until June 2010, there was no specification by product form for domestic rare-earth producers and traders. Beginning in July 2010, the MOC ceased issuing separate export quotas to domestic and Sino-foreign joint-venture rare-earth producers and ceased specification by product for joint-venture producers (China Ministry of Commerce, 2006, 2008a,b, 2009, 2010a,b,c).

Between 2005 and 2007, the Government quota allocated more than 40,000 t of rare-earth production to domestic producers and traders and about 16,000 t to Sino-foreign joint-venture producers (table 1). In 2008, the Government reduced the rare-earth export quota for domestic rare-earth producers and traders by 21.6 percent from that of 2007; in 2009, it reduced the quota by an additional 2.5 percent. The rare-earth export quota for Sino-foreign joint-venture producers remained at the level of 16,000 t in 2008 and 2009. In 2010, the Government allocated the first tranche of the rare-earth export quota, a total of 22,282 t (16,304 t to domestic and 5,978 t to Sino-foreign); this was more than the total of 21,728.1 t (15,043 t to domestic and 6,685.1 t to Sino-foreign) that was allocated for the first tranche for 2009. In the second tranche of the export quota allocated in July 2010, the Government allocated a total of 7,976 t (6,208 t to domestic and 1,768 t to Sino-foreign), compared with a total of 26,427 t (16,267 t to domestic and 10,160 t to Sino-foreign) in 2009. The rare-earth export quota in 2010 was 37.1 percent less than in 2009. In 2011, the Government established a 14,446-t rare-earth export quota (10,762 t to domestic and 3,684 t to Sino-foreign), which was about 35 percent less than the first-tranche quota for 2010 (China Ministry of Commerce, 2009, 2010a,b,c).

In addition to issuing production and export quotas, China taxes mineral production to raise revenue and to regulate the mineral industries. In the 1 990s, the Government encouraged enterprises to export their products by refunding the value-added tax that producers paid on exported products. In early 2000, owing to increased domestic consumption, the Government reduced the export rebate for many strategic commodities. In 2005, the rebate on exported rare earths was eliminated, and trade of rare-earth concentrate was banned. In 2007, the Government introduced an export duty on rare-earth products to restrict the export of products that consume large amounts of energy to produce in order to protect the domestic supply of strategic minerals (table 2).

Table 2. China's Rare-Earth Export Duty Rates

[In percentage. Abbreviation used: NA, not available. Sources: China Customs Import and Export Tariff Department (2007–2011)]

Commodity	Export duty rate (%)				
	2007	2008	2009	2010	2011
Yttrium oxide	10	25	25	25	25
Lanthanum oxide	10	15	15	15	15
Cerium oxide, hydroxide, carbonate, and others	10	15	15	15	15
Praseodymium	NA	NA	NA	NA	NA
Neodymium oxide	10	15	15	15	15
Europium and its oxide	10	25	25	25	25
Gadolinium	NA	NA	NA	NA	NA
Terbium and its oxide, chloride, and carbonate	10	25	25	25	25
Dysprosium oxide, chloride, and carbonate	10	25	25	25	25
Holmium	NA	NA	NA	NA	NA
Erbium	NA	NA	NA	NA	NA
Thulium	NA	NA	NA	NA	NA
Ytterbium	NA	NA	NA	NA	NA
Lutetium	NA	NA	NA	NA	NA
Other rare-earth oxide	10	15	15	15	15
Mixed rare-earth chlorides and fluorides	10	15	15	15	15
Mixed rare-earth carbonates	10	15	15	15	15
Mixed rare-earth, yttrium, and scandium compounds and metals (including battery grade)	10	25	25	25	25
Non-mixed rare-earth carbonates	10	15	15	15	15
Rare-earth ore	10	10	15	15	15
Metals:					
Lanthanum	NA	NA	NA	NA	25
Cerium	NA	NA	NA	NA	25
Neodymium	10	15	15	15	15
Dysprosium	NA	NA	NA	25	25
Other mixed metals	NA	NA	NA	25	25

In 2008, China exported a total of 54,963 t (gross weight) of rare-earth products, including 41,349 t of rare-earth oxides; 5,739 t of rare-earth salts; 4,550 t of rare-earth metals; 2,390 t of rare-earth alloys; and 935 t of other rare-earth products. In 2009, owing to the global financial recession, the demand for rare-earth products was weak, especially from Japan; therefore, China's rare-earth exports decreased to 43,918 t, which included 32,084 t of rare-earth oxides; 6,082 t of rare-earth salts; 3,271 t of rare-earth metals; 2,075 t of rare-earth alloys; and 406 t of other rare-earth products (Yin, 2010).

China's Rare-Earth Development Plan

According to China's draft rare-earth development plan, annual rare-earth production may be limited to between 130,000 t and 140,000 t (REO) during the period from 2009 to 2015. The export quota for rare-earth products may be about 35,000 t (REO), and the Government may allow 20 domestic rare-earth producers and traders to export rare earths. The plan was proposed to control the export of rare-earth primary products, such as rare-earth salts, and to encourage the export of new rare-earth downstream material products. The Government will not ban the export of dysprosium and terbium but will limit the volume of export for these two elements. The Government will not approve any new rare- earth separation projects before 2015. To protect rare-earth resources, rare-earth producers will be required to have a minimum mine output capacity of 300,000 metric tons per year (t/yr) of ore for light rare earths and 3,000 t/yr (REO) for ion-adsorption rare earths. The Government will ban monazite mining if the monazite contains radioactive elements. For rare-earth separation, producers will be required to have a separation output capacity of 8,000 t/yr (REO) of mixed rare earths, 5,000 t/yr (REO) of bastnaesite, and 3,000 t/yr (REO) of ion-adsorption rare earths. Metal smelting producers must have an output capacity of 1,500 t/yr.

Rare-earth producers will be required to meet the environmental emission standards; otherwise, they will be shut down (China Ministry of Environmental Protection, 2011).

The government of Nei Mongol Autonomous Region granted approval to Baotou Iron and Steel and Rare Earths Corp. to stockpile up to 30,000 t of rare-earth concentrates in Baotou. This will allow the company to increase the recovery of rare earths to 50 percent from 25 percent in the separation process (China Ministry of Industry and Information Technology, 2010; Wang, 2010).

During the past several years, the Government has continued to urge rare-earth producers to merge into large groups and to eliminate duplicate projects. Rare-earth producers realized that consolidation could benefit the rare-earth sector in China and would avoid unnecessary competition among themselves and associated financial losses. Dingnan Dahua New Materials Co. Ltd., Ganxian Hongjin Rare Earth Co. Ltd., and Minmetals Nonferrous Metals Co. Ltd. formed a joint-venture company, Minmetals Ganzhou Rare Earth Co. Ltd., to process rare-earth resources in Ganzhou, Jiangxi Province, in 2008. Most of China's ion-adsorption rare earths are located in Ganzhou. There were 88 rare-earth producers in Ganzhou, and 90 percent of them ceased their operations because of weak prices. Jiangxi Province had a reserve of 2.3 Mt of the ion-adsorption rare earths. Minmetals Ganzhou Rare Earth had a rare-earth separation capacity of 8,500 t/yr and planned to expand the separation capacity to 13,500 t/yr within 5 years. The joint venture also planned to develop value-added rare-earth products for the aerospace and electronics sectors.

Aluminum Corporation of China (Chinalco) signed an agreement with the government of Jiangxi Province to allow the company to consolidate the local nonferrous metals producers in the Province and to take shares of Jiangxi Rare Earth and Rare Metals Tungsten Group Co. Ltd. (JXTC). Chinalco and JXTC formed a joint-venture company, in which Chinalco held 51 percent of the shares, to manufacture nickel-cobalt products. JXTC was a leading producer of rare-earth products in Jiangxi Province. In 2003, Minmetals and JXTC formed a joint venture to produce tungsten products. Minmetals planned to acquire rare-earth resources in the Provinces of Guangdong and Hunan. Minmetals also planned to become the leading downstream rare-earth-product producer in the country. Chinalco and Minmetals are likely to compete for Jiangxi's rare-earth resources (China Mining and Metals Weekly, 2010).

REFERENCES CITED

[1] Chen Zhanheng. (2010). *Outline on the development and policies of China rare earth industry: Beijing*, Office of the Chinese Society of Rare Earths, April 7. (Accessed May 26, 2010, at http://www.cs-re.org.cn/rebbs/viewthread.php?tid=1706&highlight.)

[2] China Customs Import and Export Tariff Department. (2007–2011). Customs import and export tariff of the People's Republic of China: Beijing, China, Economic Daily Press, volumes are separately paged. (In Chinese.)

[3] China Mining and Metals Weekly. (2010). Chinalco, Jiangxi Rare Earth establish nickel-cobalt JV: *Interfax Information Services*, v. *9*, no. 16,. 3.

[4] China Ministry of Commerce. (2006). *Ministry of Commerce 2006 announcement of industrial export quota for foreign investors: Beijing*, China, China Ministry of Commerce Circular 107, February 7, 2 p. (In Chinese.)

[5] China Ministry of Commerce. (2008a). Ministry of Commerce 2008 first batch of general export quota of rare earths announcement: Beijing, China, China Ministry of Commerce, January 15, 2 p. (In Chinese.)

[6] China Ministry of Commerce. (2008b). Ministry of Commerce 2008 industrial export quota for foreign investors announcement: Beijing, China, China Ministry of Commerce, February 18, 2 p. (In Chinese.)

[7] China Ministry of Commerce. (2009). 2010 first batch of rare-earth export quota: Beijing, China, China Ministry of Commerce Circular 147, December 29, 2 p. (In Chinese.)

[8] China Ministry of Commerce, (2010a), 2010 first batch of industrial export quota for foreign investors: Beijing, China, China Ministry of Commerce Circular 106, February 23, 2 p. (In Chinese.)

[9] China Ministry of Commerce. (2010b). *2010 second batch of rare-earth export quota: Beijing*, China, China Ministry of Commerce, July 7, 2 p. (In Chinese.)

[10] China Ministry of Commerce. (2010c). *2011 first batch of rare-earth export quota: Beijing*, China, China Ministry of Commerce, December 28, 2 p. (In Chinese.)

[11] China Ministry of Environmental Protection. (2011). Emission standards of pollutants from rare earths industry GB 26451–2011: Beijing, China, China Ministry of Environmental Protection Standard GB 26451–2011, January 24, 13 p. (In Chinese.)

[12] China Ministry of Industry and Information Technology. (2010). *Requirement of rare earth industry entrance announcement: Beijing*, China, China Ministry of Industry and Information Technology, May 12, 5 p. (In Chinese.)

[13] China National Bureau of Statistics. (2010). *China statistical yearbook, 2009: Beijing*, China, China Statistics Press, 1,032 p. (In Chinese.)

[14] China National Material and Technical Control Bureau. (1999). *Classification for resources/reserves of solid fuels and mineral commodities GB/T 17766–1999: Beijing*, China, China National Material and Technical Control Bureau Standard GB/T 17766–1999, 9 p. (In Chinese.)

[15] China State Council. (2006). *Instruction for mineral resource development: Beijing*, China, China State Council Circular 108, December 31, 5 p. (In Chinese.)

[16] Geological Publishing House. (2010). *Rare metal and rare earth mining sector, in China mining history: Beijing*, China, Geological Publishing House, 449. (In Chinese.)

[17] Haxel, G. B., Hedrick, J. B. & Orris, G. J. (2002). Rare earth elements--Critical resources for high technology: U.S. Geological Survey Fact Sheet 087–02, 4 p., available at http://pubs.usgs.gov/fs/2002/fs087-02/.

[18] U.S. Bureau of Mines and U.S. Geological Survey (1980). Principles of a resource/reserve classification for minerals: U.S. Geological Survey Circular 831, 5 p.

[19] Wang Huachun. (2010). [The leading enterprise—Baogang Rare Earth]: *China Metal Bulletin*, no. *12*, 36–37. (In Chinese.)

[20] Wen Lu. (1998). Chinese industrial minerals—Rare earths: Surrey, United Kingdom, Industrial Minerals Information Ltd., 210 p.

[21] Yin Jianhua. (2010). [Rare earth market is expected to become better]: *China Metal Bulletin*, no. *8*, 24–25. (In Chinese.)

CHAPTER SOURCES

The following chapters have been previously published:

Chapter 1 – This is an edited reformatted and augmented verison of a United States Geological Survey publication, report 2010-5220, dated 2010.

Chapter 2 – This is an edited reformatted and augmented version of a United States Geological Survey publication, report 2011-1042, dated 2011.

INDEX

C

D

I

J

K

L

M

Q

R

S

T